E. Sheppard and Sons

Catalogue of E. Sheppard and Sons

Wholesale and retail dealers in plants for greenhouses, hothouses and bedding

E. Sheppard and Sons

Catalogue of E. Sheppard and Sons
Wholesale and retail dealers in plants for greenhouses, hothouses and bedding

ISBN/EAN: 9783741163449

Manufactured in Europe, USA, Canada, Australia, Japa

Cover: Foto ©berggeist007 / pixelio.de

Manufactured and distributed by brebook publishing software
(www.brebook.com)

E. Sheppard and Sons

Catalogue of E. Sheppard and Sons

OF

E. SHEPPARD & SONS,

Wholesale and Retail Dealers in

P
L
A
N
T
S

S
H
R
U
B
S

Fruit and Ornamental Trees, Roses, Etc.

1883.

GREENHOUSES AND NURSERY,

224 Fairmount Street . . . Lowell, Mass.

Please Read this before Ordering.

In presenting this Catalogue to our patrons, we would beg leave to return our thanks for the liberal patronage of the past, and trust to be favored with a continuance of the same.

Purchasers unacquainted with the different varieties of plants, leaving the selection with us, — stating the sum they wish to expend, and the object they wish to effect, — may depend upon receiving the best sorts and best plants.

Orders. — Parties ordering will please use the "ORDER SHEET," which will greatly facilitate their execution. If mixed up in the body of the letter, mistakes are almost unavoidable.

Persons starting in business, or florists in want of stock, will find it to their advantage to call upon us.

Shipping. — All plants sent out by us will be carefully packed and delivered in Lowell, and shipped or forwarded according to directions, after which they are at the risk of the purchaser.

We have for sale, this season, an unusually large and complete stock of Greenhouse and Bedding Plants, Evergreens, Shrubs, Vines, etc. We would earnestly request a personal inspection of the same. Those, however, favoring us with orders, may fully rely on our using every exertion to give perfect satisfaction.

Orders by mail or express will receive prompt and careful attention.

All orders from unknown correspondents must be accompanied with *cash* or *satisfactory reference.*

Hyacinths, Tulips, Tuberoses, and other Bulbous Roots imported fresh every year.

☞ *Liberal discount on large orders and to the trade.*

E. SHEPPARD & SONS,

Nurserymen and Florists,

P. O. Box 335.   224 FAIRMOUNT ST., LOWELL, MASS.

CATALOGUE

OF

E. SHEPPARD & SONS,

WHOLESALE AND RETAIL DEALERS IN

PLANTS,

FOR

Greenhouses, Hothouses, and Bedding.

GREENHOUSES AND NURSERY:

No. 224 Fairmount Street, Lowell, Mass.

New Plants for 1883.

NEW CHRYSANTHEMUMS.—(Large Flowering.)

FAUST. Crimson-purple, very fine.
INNER TEMPLE. Magenta-crimson, distinct.
L. ORIENT. Mahogany-red, fine.
MRS. PARNELL.
MRS. DIXON. Splendid yellow.
MR. JAY. Reddish-crimson, fine.
PEARL DES BEAUTIES. Deep crimson and brilliant amaranth.
PRINCE ALFRED. Rose-crimson, large and fine.

Price, 35 cts. each; $3.00 per dozen.

JAPANESE CHRYSANTHEMUMS.

BOUQUET FAIT. Rich-rose and silvery-white, centre shaded yellow, flowers large and very fine.
COMTESSE DE BOMGARD.
METRE CHINESE. Canary-yellow.
MONS. BACO. Vivid crimson and orange, with golden centre.
GILBY BLANCH.
M. PLANCHENAT. Mauve-shaded, rose, and silver.
MONS. ARDINE.
M. RICHARD LAROIS. Dark rose and bright violet, tipped pure white; an enormous flower.
SOUV. DE LA MARRIDES.
THE SULTAN. Bright rosy-purple floret, with a light back; very fine.
TRIOMPHE DU CHATELET. Salmon color, tinted rose, golden centre, immense size.

Price, 25 cts. each; $2.50 per dozen,

SINGLE DAHLIAS.

COCCINEA BUFFALO. Pale dun or fawn color, flushed with orange-red.
COCCINEA CATO. Clear orange-scarlet. The florets are yellow at the base.
CANARY. Bright canary-yellow, extra fine.
COMET. Deep reddish-scarlet, veined and suffused with orange and gold flowers; medium size, fine shape and form.
DIDO. Magenta-rose, paler at the tips, with a yellow zone at base.
GRACILIS PERFECTA. Velvety-crimson, free bloomer.
HALO. Magenta-crimson; a zone of yellow at the base of florets.
HARLEQUIN. Deep rose, ground color, having a broad band of purple down the centre of each petal; very attractive.
LOVELY.
LUTEA GRANDIFLORA. Rich yellow, large, well-shaped flower, very free, and good habit.
SCARLET GEM. Bright scarlet.
SINGLE ZINNIA. Rich crimson-scarlet; a neat flower, dwarf habit.
TYRO. Lilac-yellow zone at the base.
WHITE QUEEN. One of the best and most useful.

Price, 40 cts. each; $4.00 per dozen.

NEW SHOW DAHLIAS.

CHAMPION ROLLO. Large, dark orange, with a light shade on the edge of the petals; fine.

HON. MRS. PERCY WYNDHAM. Yellow ground, deeply edged with rosy-purple; very pleasing in color; fine flower.

H. W. WARD. Yellow ground, heavily edged and shaded with crimson; fine form.

JAMES VICK. Purple maroon, color intense, very full.

LADY WIMBORNE. Deep pink, heavily shaded with rose; very pretty; a new color.

MRS. DODD'S. Blush centre, outer petals light lilac, splendid form.

RICHARD EDWARD. Plum color, with a pretty shade of lilac on the surface; a compact and well-built flower.

WALTER H. WILLIAMS. Bright scarlet, surpassing all others of that shade; grand high centre, large size, and a profuse bloomer.

Price, 50 cts. each; $5.00 per dozen.

NEW FANCY DAHLIAS.

ANNIE PRITCHARD. White, beautifully striped with lilac-rose; a large flower.

JAMES O'BRINE. Yellow, with crimson and reddish-rose stripes; a fine flower.

LADY ANTROBUS. Red, tipped with pure white; fine form.

PROFESSOR FAUCETT. Dark lilac, striped with chocolate; a magnificent flower in every respect.

Price, 50 cts. each; $5.00 per dozen.

NEW FUCHSIAS.

GUSTAVE DORE. Rich carmine tube and sepals, double white corolla, medium size and a free bloomer.

JULES FERRY. Scarlet tube sepals, violet mottled white corolla; a pretty variety; single.

MISS LIZZIE VIDLER. Light rosy-red sepals, beautiful soft mauve corolla; very free.

MAGNUM BONUM. Sepals, broad and brilliant, red in color, large, well-formed corolla of the richest violet.

PRESIDENT. Tube and sepals bright vermilion; corolla a very rich violet; beautifully formed.

TRUMPETER.

Price, 50 cts. each; set of six for $2.50.

GYNURA AURANTIACA.

A new bedding plant. The greatest addition to our foliage plants, since the introduction of Coleus Verschaffeltii. The leaves are very large, and rich plum-purple in color, covered all over with a dense hair-like covering, which gives the plant the appearance of plush or velvet; growth vigorous.

Price, 40 cts. each.

NEW VARIEGATED LEAVED GERANIUMS.

MDME. SALLEROI. A distinct variety, with leaves from one to two inches in diameter; the centre of each is of deep olive-green, with a broad margin of pure white; habit quite dwarf; it is not affected in the least by exposure to the direct sunlight. Price, 40 cts.

MRS. PARKER. The foliage of this variety is identical with Mountain of Snow, but produces a quantity of beautiful double pink colored flowers, which renders it very distinct and attractive. Price, 50 cts.

NEW ZONALE GERANIUMS.

CARILLON. Deep salmon, of a pleasing shade, large and fine; double.

EUREKA. The finest single white; trusses large, and stand well above the foliage.

GENERAL FERRE. Orange-salmon, double flowers, large and beautifully mottled; trusses very large, and produced in abundance.

HENRI CANNELL. Deep magenta, purple large pips, full and double, highly suffused, purple.

JEAN ILL. Pink, shaded with very deep purple; immense trusses.

LIZARD. Rosy-salmon, with a distinct bright ring of red; very dwarf; flower single.

MRS. STRUTT. Purplish pink; flowers and trusses large and well-formed; a pleasing color; dwarf and free.

MRS. MOORE. Pure white, with a beautiful distinct ring of scarlet; small white eye.

MDLLE. EMILIA BERLTET. Double white.

SPENCER. Magenta, white eye; a very free bloomer.

Price, 50 cts. each.

NEW TEA ROSES.

MISS MAY PAUL. A vigorous growing variety, of climbing habit; flowers white, with lilac interior; outside petals red, large, and well-formed.

MDME. CUSIN. Purple-rose, with white centre tinted with yellow; large, full, and well-formed; vigorous growth.

Price, 50 cts. each.

NEW HYBRID TEA ROSES.

CAMOENS. A bright China rose; deep yellow, almost always striped with white; medium size.

PRINCE IMPERIAL DE BRESIL. Bright carmine-rose; large, very full, and well-formed; vigorous.

PIERRE GUILLOT. Bright dazzling red; petal bordered with white, very large, with erect flowers, well-formed; very free bloomer; growth vigorous.

REINE MARIE HENRIETTE. A red Gloire de Dijow; large and full; climbing. This superb variety is the only one in this class.

Price, 50 cts. each.

GENERAL COLLECTION.

ABUTILONS.

These plants have extremely beautiful flowers, being richly veined and striped, suitable for the garden or greenhouse; flowers bell-shaped.

AUGUSTE PASEWOLD. A fine variety, foliage beautifully marked with large patches of green, yellow, and white; very distinct.

BOULE DE NEIGE. Pure white, a most abundant bloomer; flowering freely during the winter.

DARWINI. Orange-scarlet, veined with pink.

MESOPOTAMICUM. Calyx scarlet, yellow petals.

"   VARIEGATA. Leaves variegated yellow and green.

PURPUREUM. A low-growing variety, with rosy-purple flowers, blooming in clusters.

ROSÆFLORUM. Rich rose-color, veined with crimson.

THOMPSONII. Leaves beautifully marbled with bright yellow spots.

VERSCHAFFELTII. Flowers bright yellow.

VENOSUM. Orange-red; large.

PATERSONII. Crimson, veined with maroon, of large size and great substance.

SLOWEANA MARMORATA. The leaves are very large, and beautifully mottled green and yellow; the latter predominating.

Price, 25 cts. each; $2.50 per dozen.

ACALYPHA TRICOLOR.

A very handsome plant from the Feejee Islands, with large foliage, irregularly mottled with crimson and flame-red. A beautiful plant for summer decoration. Price, 25 cts. to 50 cts. each.

A. MARGINATA. 50 cts. each.

A. MACAFEANA. 50 cts. each.

ACHYRANTHUS.

Fine plants for vases and bedding out; particularly recommended for ribbon gardening, etc.

BRILLIANTISSIMA. A most brilliant variety, with heart-shaped leaves of a brilliant carmine mottled with crimson, the stem and leaf-stalks pinkish-violet.

LINDENII. Fine dwarf habit, lance-shaped leaves, of a deep blood-red color.
LINDENII BRILLIANTISSIMA. Leaves and habit the same as A. Lindenii; but with deep carmine leaves, and deep pinkish-violet stems.
LINDENII AUREA. Leaves bright yellow, veined carmine; stems violet-crimsom.
VERSCHAFFELTII. Deep crimson leaves, with bright carmine veins.
AUREUS RETICULATUS. Leaves light green, marked with net-work of golden yellow.

Price, 10 cts. each; $1.00 per dozen.

AMPELOPSIS VEITCHII.

A small variety of the Virginia Creeper; a plant of rapid growth, and adheres very firmly to any surface. It is perfectly hardy, and, like the old variety, the leaves change to a bright scarlet in the fall.   Price, 25 cts. each.

ARDISIA CRENULATA. Red berries; 50 cts.
ACACIA. In variety; 50 cts.
ACORUS GRAMINEUS. Foliage variegated; 25 cts.
ALOE VARIEGATA; 50 cts.

ALTERNANTHERAS.

Beautiful dwarf-growing plants from Brazil; finely variegated with all the shades of crimson, scarlet, yellow, and green; suitable for baskets and ribbon-lines.

PARONYCHIOIDES. Leaves tinted green, crimson, and straw-color.
PARONYCHIOIDES, MAJOR. Leaves brilliant carmine and green.
PARONYCHIOIDES, MAJOR AUREA. Bright golden yellow, retaining its color throughout the season.
SPATHULATA. Leaves tinted carmine and green.
VERSICOLOR. Leaves tinted light rose to deep crimson.

Price, 10 cts. each; $1.00 per dozen.

AKEBIA QUINATA.

A beautiful *hardy* climber, attaining a height of twenty feet. Flowers chocolate-purple color; very fragrant. Price, 25 cts. each.

ARALIA FILICIFOLIA.

An ornamental stove plant, graceful in habit, and well furnished with foliage. One of the most valuable decorative plants of its family. The stem and leaf stocks are purplish, thickly marked with oblong white spots, with leaves expanding into a broad leafy limb which is imparipinnately divided. Price, 75 cts. each.

ARALIA SIEBOLDII.

A beautiful ornamental plant, alike useful in the conservatory or the sub-tropical garden; leaves palmate, deeply lobed, bright green. Price, $1.00 each.

AGERATUM MEXICANUM VARIEGATUM.

An easily grown garden favorite, of compact habit; leaves finely variegated yellow and green, shaded with pink; good bedder.

A. MEXICANUM. Blue flowers.

A. PRINCE ALFRED. A dwarf variety of A. Mexicanum; flowers of a delicate pink shade.

Price, 10 cts. each; $1.00 per dozen.

ALOYSIA CITRIODORA.

Well known as the "Lemon-Scented Verbena." Indispensable in every garden, for the delightful fragrance of its leaves. Price, 15 cts. each; $1.50 per dozen. .

ASTILBE JAPONICA. — (Spiræa Japonica.)

One of the most beautiful of all hardy herbaceous plants, growing about eighteen inches high, with foliage of a dark glossy green, flowering in spikes of pure white feather-like flowers· a valuable plant for winter flowering. Price, 25 cts. each; $2.50 per dozen.

ANEMONE HONORINE JOBERT.

A beautiful herbaceous plant; producing large white flowers from August to October; grows two to three feet high. Price, 35 cts. each.

ANTHERICUM VITATUM VARIEGATUM.

A useful and beautiful variegated-leaf plant; one of the best for hanging basket. Price, 25 cts. each; $2.50 per dozen.

AGAVE AMERICANA VARIEGATA. — (Century Plant.)

No plants are more decorative or more effective than these; they require but little attention, and may be wintered under the stage of the greenhouse, or in a dry, warm cellar. Price, 50 cts. to $1.00 each.

AZALEA INDICA.

We have a large variety of this valuable species, too numerous to describe, consisting of all the leading varieties. Price, 50 cts. to $3.00 each.

HARDY AZALEAS.

Of all the hardy flowering shrubs, none, perhaps, afford such a variety in color as the Azaleas, for almost every shade of pink, white, yellow, orange, and scarlet is to be found among them; and as they generally flower in great profusion, and are, many of them, deliciously scented, they deserve to be universally planted. The following varieties are perfectly hardy, and will flourish wherever Rhododendrons are grown.

2

AMŒNA. White and pink, fine.
BESSIE HOLDAWAY. Pink, fine.
CARDINAL. Bright pink.
CARDONIANA. Pink, light shade.
CHAS. BAUMANN. Fine, pink.
GLORIA MUNDI. Bright red and orange.
JULES CÆSAR. Orange and pink.
MACRANTII. Light yellow.
MADAME BAUMANN. Bright pink, striped with white and orange.
MELANIE. Light rose, fine.
METEOR. Orange and brick.
MINERVA. Pink and orange.
NANCY WATERER. Bright yellow, fine.

CALENDULACEA ELEGANS. Pink, orange, and white.
CUPREA. White, shaded pink.
FAMA. Rosy pink.
FULGENS. Salmon-pink.
OPTIMA. Bright orange.
PALLAS. Crimson and orange.
PRINCEPS. Fine yellow.
PULCHELLA ROSEOLA. Pink.
RADIATA. Pink and orange.
ROI DES BELGES. Rosy pink, striped with white.
SINENSIS ROSEA. Yellow.
TRIUMPHANS. Crimson-pink.
UNIQUE. Fine orange.
VISCOCEPHALA. Pure white, extra.

Price, $1.50 to $3.00 each.

BOUVARDIA ALFRED NEUNER.

A new double white variety. This charming novelty will prove of inestimable value for all kinds of decorative purposes; it is of excellent habit, and a profuse bloomer, throwing large trusses of lovely pure white rosette-like flowers, each flower composed of three perfect rows of petals. Where cut-flowers are required, this beautiful plant is unequalled. Price, 50 cts. each; $4.50 per dozen.

BOUVARDIAS.

Splendid wintering-flowering and bedding plants. The following varieties are the most desirable:—

ELEGANS. Bright carmine, large truss.

DAVIDSONII. White, fine form.
LEANTHA. Dark dazzling-scarlet.

Price, 25 cts. each; $2.50 to $5.00 per dozen.

BEGONIA GLAUCOPHYLLA SCANDENS.

A drooping or creeping variety, with large clusters of rich salmon-colored flowers; very desirable for hanging-baskets. Price, 25 cts. each.

BEGONIA.

We have a large collection of this popular plant. The class of which B. Rex is a type, is extensively used for baskets, vases, ferneries, etc. Of the winter-flowering varieties, B. Fuchsloides, B. Weltoniensis, are representatives, producing a profusion of pink, white, and carmine flowers during the entire winter months. Price, 25 cts. each; $2.50 to $4.50 per dozen.

BEGONIAS.—(Tuberous Rooted.)

We have a fine collection, consisting of some of the best named varieties. In June and July we shall have a large lot of seedlings saved from the best

named sorts. Price, named varieties, 50 cts. to $1.00; seedlings, 35 cts.
each ; $3.50 per dozen.

BEGONIA DIAMANT.

A beautiful variety of the Rex type, with medium-size foliage, of a rich
silvery huc tinted with rosy-pink, dwarf habit. Price, 25 cts. each.

BEGONIA METALLICA.

Perhaps of all the ornamental-foliaged Begonias, none are so beautiful as
this, and for a window or drawing-room plant, it is without a rival. Price,
35 cts. each.

BEGONIA RUBRA.

One of the finest winter-flower variety. The flowers are scarlet-rose,
glossy, and wax-like, very large and freely produced. Price, 25 cts. each ;
$2.50 per dozen.

BOUGAINVILLEA SPECTABILIS.

A handsome, free-flowering greenhouse climber, of very rapid growth ;
flowers rich mauve color. Price, 50 cts. each.

BOUGMANSIA SUAVEOLENS.

A very showy and beautiful plant, growing from three to six feet high,
with large, drooping, trumpet-shaped flowers ten inches long, white and very
fragrant; blooming profusely all summer. Price, 50 cts.

CALADIUMS.

A fine collection of these beautiful, ornamental-foliaged plants for the
decoration of the conservatory or greenhouse in summer, with large-sized
leaves assuming almost every imaginable color in their variegation. They
are of easy culture. Many of the varieties are well adapted for baskets,
vases, etc. Price, 50 cts. to $1.00 each.

ADOLPHE ADAM. Densely speckled with white and red.
ARGYRITES. Spotted profusely with white.
BICOLOR. Dark crimson mid-rib.
CHANTINI. Profusely spotted carmine, crimson mid-rib.
DEVOSIANUM. Dotted and flecked with pure white.
DUCHARTRE. White, suffused with rose.
HOULETTII. Light green, white spots, pink mid-rib.
MDME. HOULETTII. Pink and white spots.
MARS. Large ; crimson centre.
PICTUM. Light green ground, large white spots.
REINE VICTORIA. White; green and crimson spots.

Price, 50 cts. to $1.00 each ; $4.50 to $10.00 per dozen.

CAMPANULA MARITIMA.

A handsome, blue-flowering plant, of trailing habit; it flowers freely; very effective as a basket-plant. Price, 25 cts. each; $2.50 per dozen.

CANNAS.

These plants, by their broad, massive foliage, impart a beautiful aspect to gardens; also well adapted for pot-culture. Price, 25 cts. each; $2.50 per dozen.

CALCEOLARIA — SHRUBBY.

BIJOU.   Brilliant crimson, dwarf.
GOLDEN GEM.   Deep yellow, profuse bloomer.

Price, 10 cts. each; $1.00 per dozen.

COLOCASIA ESCULENTA.

One of the most attractive ornamental-foliaged plants in cultivation. The leaves are heart-shaped, of immense size. As a single plant for the lawn or large flower-borders, it has no superior. Price, 25 to 50 cts. each; $2.50 to $5.00 per dozen.

MONTHLY CARNATIONS.

ADONIS.   Variegated, red and white.

ASTORIA.   Yellow, scarlet, and white.

FRED JOHNSON.   Brilliant scarlet, very fragrant.

GRACE WILDER.   A beautiful clear pink, finely fringed.

GEN. GRANT.   Pure white, in clusters.

HENRIETTA.   Striped, rose and purple.

LA PURITE.   Rosy-carmine, profuse bloomer.

MISS JOLIFFE.   A delicate blush-pink.

PRINCESS LOUISE.   Bright carmine rose; quite fragrant; free bloomer.

PRESIDENT DEGRAW.   Pure white, fragrant.

SMITH'S SEEDLING.   Large, white, fragrant.

VARIEGATED LA PURITE.   Carmine, striped.

Price, 15 cts. to 25 cts. each; $1.50 to $2.50 per dozen.

CENTAUREA CANDIDISSIMA.

A valuable plant for massing, or to contrast with dark-colored foliage in ribbon-rows. Leaves downy-white; compact habit. Price, 20 cts. to 50 cts. each; $2 00 per dozen.

CENTAUREA GYMNOCARPA.

A beautiful plant with silvery-gray foliage; well adapted to contrast, in ribbon lines, with dark-foliaged Coleus or Achyranthus. As a basket-plant it is unsurpassed, its drooping, fern-like leaves being very effective. Price, 10 cts. each; $1.00 per dozen.

CERASTIUM TOMENTOSUM.

A white-foliaged plant of trailing habit; well suited for hanging-baskets or stands. Price, 10 cts. each; $1.00 per dozen.

COROZEMA VARIA.

A beautiful greenhouse shrub, bearing purple and orange-colored flowers in spikes four to six inches in length, lasting in bloom through January and February. Price, 35 cts. each.

CHRYSANTHEMUMS.—(Large Flowering.)

ALARM. Light pink, extra.
ANGELINA. Bronzy yellow.
BOADICEA. White, tinted pink.
BRONZE JARDIN DES PLANTES. Bronze-orange; yellow centre.
DUKE OF ROXBURGH. Bright yellow.
DUCHESS OF MANCHESTER. White, large flower.
EMPRESS OF INDIA. Clear white.
ELLEN TURNER. Lilac-rose.
GOLDEN PERFECTION. Bright golden-yellow.
GOLDEN QUEEN. Golden-yellow.
GEN. HALLEY. Light rose.
GENICE. Rosy pink.
GUERNSEY NUGGET. Primrose-yellow.
JARDIN DES PLANTES. Golden-yellow.
LADY SLADE. Delicate lilac, pink-blush centre, incurved flower.
LORD STANLEY. Buff.
LA PURITE. Creamy white.
M. LUCIAN BALK. Reddish crimson and orange.
MRS. HALIBURTON. Sulphur-white.
MISS MARY MORGAN. Delicate pink, fine.
MRS. GEO. RUNDEL. Pure white, incurved flower.
NE PLUS ULTRA. Rose.
PRINCE OF WALES. Dark purple-violet.
PROGNE. Bright crimson.
QUEEN VICTORIA. White.
RIFLEMAN. Ruby-red, finely incurved.
ROSE PERFECTION. Bright rose.
ROSE TRAVENNA. Pink.
SUNFLOWER. Pale canary color.
SNOWBALL. White, ball shape.
VOLUNTEER. Rosy-lilac.
VENUS. Delicate peach.
WHITE VENUS. White, finely incurved.
WHITE CLOUD. Snow-white.

Price, 15 cts. to 25 cts. each; $1.50 to $2.50 per doz.

JAPANESE CHRYSANTHEMUMS.

CHANG. Yellow, tipped with crimson.
CHINESE PURPLE. Purple.
DR. MASTERS. Centre bright red, tipped with gold.
ERECTA SUPERBA. Rich, rosy purple.
FAIR MAID OF GUERNSEY. Pure white, extra large.
HENRY BROCK. Purple-crimson.
HELEN McGREGOR. Orange and crimson, small.
JAMES SALTER. Pink, changing to white.
JONAS.
LA NEGRO. Very dark maroon.
MT. ETNA. Orange, tipped, or shaded crimson.
MAJESTA VIOLET. Violet-purple, spotted white.
ORPHEE. Brick-red and crimson, centre gold, incurved flower.
THE COSSACK. Golden-yellow and crimson.

Price, 15 cts. to 25 cts. each; $1.50 to $2.50 per doz.

POMPONE CHRYSANTHEMUMS.

ANNA DU BELOCCA. Sulphur-white.
ADMIRAL.
ADAM FORSYTH.
AMPHILLIA.
DANTON. Brick-red and orange.
DURUFLET. Rose-carmine.
DAMIETTE. Blush, dark centre.
GEN'L CANROBERT. Pure yellow.
GOLDEN CIRCLE. Golden-yellow, quilled flower.
JUSTIN TESSIER. Blush-white.
LILAC CEDO NULLI. Lilac.
LA FIANCEE. White, fringed.
LITTLE HARRY.

MODEL OF PERFECTION. Rich lilac edge, pure white, distinct.
MRS. CAMPBELL.
MELESE. Dark-maroon.
MADAME DE VATRAY. Rosy-lilac.
MADAME DOMAGE. White, early, fine flower.
MARIA. Lilac rose.
MAROON MODEL. Maroon-crimson.
NELLIE. Creamy-white.
PRINCE OF LILIPUTIANS. Yellow.
ROSE MANTLE. White, pink shaded.
SCARLET GEM.
SMALL BRONZE. Bronze.

Price, 10 cts. to 25 cts. each; $1.00 to $2.50 per dozen.

CINERARIA MARITIMA CANDIDISSIMA.

A beautiful silvery-foliaged plant, compact habit. Price, 10 cts. each; $1.00 per dozen.

COBŒA SCANDENS.

A well-known climbing plant of vigorous growth, and may easily be trained to a height of thirty feet if desired. Price, 25 cts. each; $2.50 per dozen.

COBŒA SCANDENS VARIEGATA.

A beautiful variety of the above. The leaves are margined with yellowish-white, which form a beautiful contrast to its large, purple, bell-shaped flowers. Price, 25 cts. each.

COPROSMA BAUERIANA VARIEGATA.

An exceedingly handsome plant, with bright, glossy, green-and-yellow leaves; excellent for decorating; also a good bedder. Price, 35 cts. each.

CYCLAMEN PERSICUM.

A fine collection of these beautiful winter-blooming plants, comprising all the shades of color from pure white to deep crimson, many of the varieties being striped and blotched. Price for good, strong, flowering plants, 50 cts. each; smaller plants, 25 cts. each; $2.50 per dozen.

CISSUS DISCOLOR.

A handsome hothouse climber, with leaves beautifully shaded with green, purple, and white. It should be grown in a moist, warm atmosphere, and partially shaded from direct rays of the sun. Price 25 cts. each; $2.50 per dozen.

COLEUS.—(General Collection.)

The Coleus are now too well known to require any description. Valuable for massing or ribbon borders, in contrast with lighter foliage. We have selected the following as the most distinct of the named varieties.

ASA GRAY. Orange, crimson, violet veins and centre; dark green maculation, green, serrated edge.

BEACON. Blackish purple, with brown and bright crimson mid-ribs and veins; foliage large.

BISMARCK. Rich, velvety, dark crimson centre, golden-yellow edge; dwarf.

DR. JOE HOOKER. Dark crimson, stained with dark brown; narrow, dark green margin.

ECLIPSE. Scarlet, shaded with brown; yellowish-green serrated edge.

EXCELSIOR. Deep yellow, slightly stained with green, and maculated with crimson.

GEO. SIMPSON.

GOLDEN CIRCLE. Rich bronze-crimson centre; broad yellow margin; one of the best.

GOLDEN GEM. Crimson-bronze, margined with bright sulphur-yellow.

HIAWATHA. Orange, yellow, and crimson; flamed with deep rich crimson.

HOVEY'S SEEDLING. Centre purplish-red; broad yellow margin.

IDYL. Veined and mottled with green and yellow, and stained with bronze red.

JAMES BARNSHAW.

KENTISH FIRE. Centre deep crimson, bronze zone; green edge.

LORD FALMOUTH. Centre pale yellow, zoned with carmine; yellowish green edge.

MARVELOUS. Brilliant crimson, marbled yellow and intense brown.

MISS RETTA KIRKPATRICK. Large white centre, shaded with yellow; broad green, lobed margin; large foliage.

MISS ROSINA.

M. J. LINDEN. Bronzy-crimson centre; broad, bright yellow edge.

NELLIE GRANT. Light-crimson centre, deep yellow edge, distinct.

PAROQUET. Yellow, maculated with crimson and green

PHARO. Rich crimson scarlet, mottled with yellow; blackish margin.

PRINCEPS. Dark crimson, stained with rich velvety brown; yellowish-green margin.

RED CLOUD. Rich crimson, evenly marmorated with blackish-brown; narrow green margin.

SERAPH. Fiery crimson, spotted with chocolate, bright green edge.

SUPERBISSIMA. Blackish maroon, with a brilliant, broad purple band through the centre of the leaf.

THOS. MEEHAN. Dark carmine, shaded with brown; with oak-leaf-shaped foliage.

VERSCHAFFELTII. Rich dark crimson, finest bedding sort.

ZEPHYR. Rich, bronze crimson, slightly marbled with dark olive-green; violet-purple veins.

Price, 10 cts. to 15 cts. each: $1.00 to $1.50 per dozen.

CLEMATIS. — (Virgin Bower.)

The Clematis are beautiful, free-going, fast-climbing, and perfectly hardy vines. They are well adapted for training on trellis-work, and grow from ten to fifteen feet high.

ALBERT VICTOR. Deep lavender, with brown rib changing to white.

BEAUTY OF SURREY.

COUNTESS OF LOVELACE.

FAIR ROSAMOND.

FLORIDA. White.

FLAMULA. White, sweet-scented.

GIPSY QUEEN.

GEM. Deep lavender-blue.

JACKMANII. Violet-purple, veined centre.

JOHN GOULD VEITCH. Light blue, large and double.

JOHN MURRAY.

LANUGINOSA NIVEA. Pure white, fine.

LORD LONDESBOROUGH. Deep mauve, purplish-red band.

LUCIE LEMOINE. Double, white, extra.

LADY STRATFORD DE RADCLIFFE.

LORD MAYO.

MRS. QUILTER.

MRS. JAMES BATEMAN.

MRS. MELVILLE.

MISS BATEMAN. Fine white.

MRS. S. C. BAKER. French-gray, one of the best.

PURPUREA ELEGANS.

RUBRO VIOLACEA. Reddish-violet.

RUBELLA. Beautiful, rich claret-purple.

STELLA.

SIR GARNET WOLSELEY.
THE QUEEN.
TUNBRIDGENSIS. Dark blue, shaded with purple.
VESTA.

Price, 50 cts. to $1.00 each.

CROTONS.

Handsome hothouse plants, with very ornamental foliage.

ANGUSTIFOLIA.	MOOREANUS.
ACUBAFOLIA.	MULTICOLOR.
AUREA MACULATA.	NOBILIS.
CORNUTUM.	PRINCE OF WALES.
DISRAELI.	PICTUM.
ELEGANS.	QUEEN VICTORIA.
EVANSIANUS.	VARIEGATUM.
FASCIATUS.	VOLUTUM.
INTERRUPTUM.	VEITCHII.
IRREGULARE.	WEISMANII.
MACARTHURII.	YOUNGII.
MAXIMUM.	

Price, 50 cts. to $2.00 each.

CURCULIGO RECURVATA.

A noble plant of Indian origin. The leaves are long, stalked, spreading, lanceolate, longitudinally plated leaves, beautifully recurved, extremely useful for decorative purposes. Price, 50 cts. to $2.00 each.

CURCULIGO RECURVATA VARIEGATA.

This is a remarkably handsome and ornamental stove-plant, producing, from a tuberous rhizoma, an arching head of recurved, plated, oblong-lanceolate leaves upwards of two feet long and six inches wide, on stalks a foot and a half in length. The leaves are green, banded in a varying manner, with clear white stripes. Price, $3.00 to $5.00 each.

CYPERUS ALTERNIFOLIUS VARIEGATUS.

A beautiful, grass-like plant, throwing up stems to the height of three feet, surmounted by a cluster or whorl of leaves. The stems and leaves are beautifully variegated with white; splendid plant for aquariums, fountains, baskets, etc. Price, 50 cts. each.

SHOW DAHLIAS.

Our varieties of this most beautiful and showy fall flower have been selected with special reference to constancy and continuance of bloom. They are all first-class varieties. Many of the new and high-priced varieties of last season are included in this selection.

3

ANNIE NEVILLE. White, shaded lilac.
BOB RIDLEY. Red, good centre.
BOMBE DE SEBASTOPOL. Large, deep scarlet.
CHRISTOPHER SCHMIDT. Light salmon, fine form.
CHAS. LIDGARD. Deep yellow, edged with red.
CHAS. BACKHOUSE. Scarlet, fine bright color.
CHAS. LIECASTER. Beautiful, bright scarlet.
CONSTANCY. Yellow ground, deeply edged with lake, a telling flower
DELICATA. Rose-fawn.
DR. ROZIES. Bright scarlet, fine form.
HENRY BOND. Bright rosy-lilac, full size; a grand flower.
JOHN BENNETT. Yellow, deeply edged with scarlet.
JOHN McPHERSONS. Violet-purple, fine form.
LAURA LIVINGSTON.
LOUISA NEAT. Delicate pink, creamy-white centre; splendid form.
MODEL. White, full centre.
MR. J. C. REID. Light orange; large, well-built flower; very constant.
MRS. HODGSON. Yellow ground, heavily edged with crimson; good petal; centre and outline grand.
MRS. SEAMAN. Yellow ground, edged with lake.
NELLIE BUCKLE. Lilac-shaded rose, full and fine.
PAUL OF PAISLEY. Lilac.
PRE-EMINENT. Purple or plum-color, extra.
QUEEN OF BEAUTIES. Pale straw, beautifully tipped with purple; fine form.
TRIOMPHE DE ROUBAIX. Rosy-amber, pointed with white.
WALTER REID. Purple, with magenta tinge, fine flower, full and large.

Price, 25 cts. to 50 cts. each; $2.00 to $3.00 per doz.

POMPONE, OR BOUQUET DAHLIAS.

DR. SCHWABES.
FANNY WEINER. Yellow, with a light crimson edge.
FRED MULLER. Reddish buff.
FIRE-BALL. Bright orange-red.
FLORA McDONALD. Primrose.
FLORIBUNDA. Rich carmine-red.
GEM. Crimson.
GEM OF THE DWARFS. Purple-crimson, tipped with white.
GOLD METEOR. Golden-yellow.
KARL GOLDENBURG. Small yellow, distinctly tipped with white.
KELEINE MOCLETTIN. Maroon.
KELEINER SERIES. Scarlet.
LITTLE DEAR. Blush-white, tipped with rose.
    "   ELIZABETH. Rosy-lilac, tipped.
    "   GOLD-LIGHT. Creamy-white.
    "   HELEN. Light blush.
    "   KATE. Velvety-purple.
    "   WONDER. Crimson-scarlet.
LURLINE. Primrose-yellow.

LINDA. Orange.
PEASANT GIRL. White, edged with crimson.
PRINCE OF LILIPUTIANS. Maroon.
PURE LOVE. Lilac.
RUBINCENTIFLORA. Dark maroon, almost black, scarlet tip.
ROUGIER CHAUVERIARE. Light blush.
SACRAMENTO. Yellow, edged with red.
SAPPHO. Dark crimson.
SCHMITZ' DEFIANCE. Yellow, white tip.
SERAPH. Orange, shaded.
VIVID-FLORA. Green flowers, very curious.
YELLOW PET.

Price, 20 cts. to 35 cts. each; $2.00 to $3.00 per doz.

FANCY DAHLIAS.

AMERICA. Rosy-lilac, striped and mottled with white.
FLORENCE STARK. White ground, striped with purple.
GOLDEN EAGLE. Yellow, fine, laced with purple.
JESSIE MCINTOSH. Red, distinctly tipped with white.
PROSPERO. Crimson, tipped with purple.

Price, 25 cts. to 35 cts. each; $2.00 to $3.00 per doz.

BEDDING DAHLIAS.

FLAG OF TRUCE. White, flaked lilac.
MT. BLANC. White; free-flowering.
QUEEN VICTORIA. Fine yellow.

Price, 25 cts. each; $2.50 per doz.

DAHLIA JAUREZII.

A most valuable and useful decorative plant for the late summer and autumn months. Its blossoms are of a rich crimson, and very much resemble, in shape and color, the well-known cactus, "Cereus Speciosimus." Height about three feet; very bushy; flowers of very striking appearance, quite unlike those of any ordinary double dahlia, the florets being flat and not cupped. Price, 25 cts. each; $2.50 per doz.

SINGLE DAHLIAS.

AURANCIACA.
LUTEA. Bright yellow.
GRACILIS.
PARAGON. Dark maroon.
SCARLET CERVANTESII. Bright scarlet.

Price, 25 cts. each; $2.50 per doz.

DAISIES.

A fine collection of these handsome, spring-flowering plants, of various colors.

BLUSH.
OLD RED. Quilled.
RED. Quilled, foliage variegated.
(See Bellis Perennis Maculata.)

RED. Quilled, very large.
WHITE.
WHITE. Quilled.
WHITE. Tipped with pink.

Price, 10 cts. each; $1.00 per doz.

DRACÆNA SUPERBA.

A robust-growing variety, having leaves fifteen to twenty inches long, by one and a half wide; beautifully recurved; dark bronzy red, changing to bright crimson. Price, $1.00 each.

DRACÆNAS.

Beautiful ornamental-leaved plants, much used as centre plants in rustic baskets, vases, etc., and are among the most beautiful plants for table decoration.

AMABILIS. Ground color bright, shining green, suffused with creamy-white and rosy-pink.
BAPTISTII. Light green foliage, streaked with bright rose.
COOPERII. Large, purple-and-red foliage.
CONGESTA. Narrow, green leaves, rapid growth.
FERREA. Broad, dark-brown leaves.
GRACILIS. Green foliage, of a graceful habit, with drooping leaves.
GUILFOYLEI. Bright green foliage, striped pink and white.
HENDERSONII. Distinctly striped with white and rosy-pink.
HYBRIDA. Deep green, margined with rose.
IMPERIALIS. Dark ground, changing to bright red and light rosy-pink.
MACLEAYI. Very dark bronze, with metallic lustre; dwarf.
MAGNIFICA.
MOOREANA. A bright reddish-crimson color, of graceful habit.
NOBILIS. High-colored variety, leaves greenish-bronze.
NOBILIS STRICTA. Marked with crimson.
PORPHYROPHYLLA. Deep bronzey-red hue on the upper side, and slightly glaucous beneath.
RUBRA ELEGANS. A narrow-leaved variety, striped red, margined with rose.
SHEPPERDII. Young leaves dark green, striped with paler green, which changes with age to orange-red.
TERMINALIS. Bronze-red, changing to bright crimson.
VEITCHII. Narrow green-leaved variety, of graceful habit.
YOUNGII. Young leaves light green, tinged with rose, changing to copper color.

Price, 50 cts. to $1.00 each.

DIEFFENBACHIA BAUSEI.

A stalky-growing variety, with broad leaves, of dwarf habit, color of leaves a yellowish-green hue, irregularly edged and blotched with dark green, and spotted with white. Price, $1 50 each.

ECHEVERIAS.

This beautiful and interesting genus of succulent plants is now attracting unusual attention. They are desirable as pot-plants for decorative purposes.

ATROPURPUREA. Long, purplish leaves.
METALLICA. A very ornamental plant, with large, massive leaves of a beautiful metallic hue.
RETUSA SECUNDA.
SANGUINEA MINIATA NUDA.
SECUNDA GLAUCA. A small, compact-growing variety, with green leaves.

Price, 10 cts. to 50 cts. each.

ERANTHEMUM TRICOLOR.

This plant requires a warm temperature to bring out its variegations, which are pink, maroon, and purple. Valuable for massing. 25 cts. each.

ERANTHEMUM ATROPURPUREUM.

A beautiful variety, with rich, rosy-purple leaves, requiring the same treatment as E. Tricolor. 25 cts. each.

ERIANTHUS RAVENNÆ.

A tall, ornamental, reed-like grass, having a beautiful foliage and dense, silvery-white plumes. Being perfectly hardy it will prove a most desirable plant for the decoration of lawns. Price, 25 cts. each; $2.50 per dozen.

EUONYMUS.

These beautiful, dwarf, glossy-leaved shrubs, by their lovely variegated foliage, have a striking and pretty appearance not possessed by any similar plant; suitable for hanging-baskets, vases, etc.

E. JAPONICA AUREA VARIEGATA.   E. RADICANS VARIEGATA.
E. SULPHUREA VARIEGATA.   E. RADICANS ROSEA VARIEGATA.

Price, 30 cts. each; $3.00 per dozen.

EULALIA JAPONICA VARIEGATA.

A very ornamental grass, of easy culture, perfectly hardy. The leaves are long and narrow, and striped green-and-white. The flower-stem is from four to six feet high, terminating with a head or panicle of blossoms, which, when cut off, retain their beauty a long time. Price, 50 cts. to $1.00 each.

EUCHARIS AMAZONICA.

A very handsome greenhouse plant. The flowers, which are freely produced, are star-shaped, pure white, and deliciously fragrant; four inches across. Easily grown in a warm, moist atmosphere. Price, $1.00 to $2.00 each.

EXOCHORDA GRANDIFLORA.

A beautiful, hardy shrub, growing about four feet high; producing beautiful spikes of white flowers in May and June. Price, $1.00 each.

FUCHSIAS.

AVALANCHE. Bright carmine sepals, violet corolla, flowers double.

BO-PEEP. Large, single, scarlet sepals, violet corolla.

BRILLIANT. Corolla scarlet, sepals white.

BOLIVIANA. An ornamental and attractive species; its branches terminating in large bunches of beautiful drooping flowers three inches long, of a rich carmine-crimson.

BERENICE. Tube and sepals rich crimson; the sepals broad and beautifully recurved; corolla deep purple, marked with rosy crimson at the base.

CLARINDA. Huge-spreading double white corolla, which is exceedingly attractive; short tube and broad sepals of a rich crimson.

CREUSA. Rich crimson tube and sepals, the latter short and completely reflexed; large corolla, a rich dark purple plum-color, shaded crimson at the base.

CARMINATA. Tube and sepals carmine, the latter very broad and well reflexed, deep violet corolla; the petals are very round and smooth.

CHARMING. Scarlet tube and sepals, purple corolla.

CLIPPER. Tube and sepals carmine-scarlet and well reflexed, corolla deep purple, shaded violet.

COCCINIA. Scarlet, self-color.

EARL OF BEACONSFIELD. Tube and sepals rosy-carmine, corolla deep crimson; free-blooming variety.

ETHEL. Tube and sepals white, rosy-violet corolla.

ECLIPSE. Large corolla of a deep purple, fine shape; bright red tube and sepals, latter broad and well reflexed; good habits; free-flowering.

FAIREST OF THE FAIR. Sepals and tubes white, rich velvet corolla.

GRAND DUCHESS MARIA. White tube and sepals, rose corolla; strong grower.

GRAND DUCHESS. Large, pure white, double corolla, of fine form; brilliant carmine tube and sepals, latter well reflexed; good habit, free-flowering. One of the best and largest double whites.

LEAH. Very long tube and sepals and corolla, tube and sepals white, corolla violet-pink; one of the best light fuchsias.

LYROS.

LYE'S FAVORITE. White tube and sepals, rich magenta corolla; flowers large and of fine form, borne in elegant clusters; good habit.

MARKSMAN. Double, blue corolla.

MARQUIS DE BELLFONT. Sepals crimson, corolla violet.

MRS. H. CANNELL. Scarlet tube and sepals; large, double, pure white corolla, the finest double white.

MONSTROSITY. The tube of this flower is rather short and stout, also the sepals, and of great thickness, and instead of reflexing in the usual way, peculiarly clasp the corolla, which is purple, shaded with violet.

NOVEAU MASTIDONTE. Excellent habit, and of free growth; flowers full and double; the corolla globular, color dark violet, veined with red; sepals beautifully reflexed.

ORANGE ROVER. Golden leaves, tinted with bronze.

PEERLESS.

PRINCESS BEATRICE. The tube is rather stout and short, sepals deep crimson, corolla reddish purple, good-shaped flower.

ROSE OF DENMARK. Tube and sepals blush, the latter beautifully recurved, corolla delicate pink.

ROSE OF CASTILE. Blush-white sepals, corolla rosy-purple.

THERA. Short tube and horizontally-reflexed sepals, of a light carmine-rose; fine large corolla, of a deep purplish plum-color.

VAINQUEUR DE PUEBLA. Sepals bright red, corolla white.

VENUS DE MEDICI. Tube white, sepals blush-white, corolla blue.

Price, 25 cts. each; $2.50 per doz.

FERNS.

No plants are more general favorites than ferns. Their great diversity and gracefulness of foliage make them much valued as plants for ferneries, baskets, rock-work, etc.

ADIANTUM AFFINE.
" AMABILE.
" CUNEATUM.
" FARLEYENSE.
" FORMOSUM.
" MACROPHYLLUM.
ANEMIA FLEXUOSA.
ASPLENUM LONGISSIMUM. Price, 75 cts. each.
CHEILANTHES ELEGANS.
DAVALLIA CANARIENSIS.
" MOOREANA.
" TENUIFOLIA.
DOODIA CAUDATA.
GYMNOGRAMMA CHRYSOPHYLLA.
" PERUVIANA ARGYROPHYLLA.
GYMNOGRAMMA TARTAREA.
LOMARIA GIBBA.
LYGODIUM SCANDENS.
MICROLEPIA STRIGOSA.

NEPHRODIUM MOLLIE CORYMBIFERUM.
NEPHROLEPIS EXALATA.
" DAVALLIOIDES FURCANS.
PHLEBODIUM AUREUM.
PLATYCERIUM ALCICORNE.
PLATYLOMA ROTUNDIFOLIA.
POLYPODIUM VACINIFOLIUM.
POLYSTICHUM PROLIFERUM.
PTERIS ARGYREA.
" CRETICA.
" " ALBA-LINEATA.
" HASTATA.
" LONGIFOLIA.
" SCABERULA.
" TREMULA.
" TRICOLOR.
SELAGINELLA CIRCINALIS.
" CÆSIUM-ARBOREA.
" DENSA.
" FLABELLATA.

Selaginella Japonica.	Selaginella Rubricaulis.
" Kraussiana.	" Umbrosa.
" Martensii.	" Robusta.
" Poulterii.	" Wildenovii.

Price, 15 cts. to 50 cts. each; $1.50 to $5.00 per doz.

TREE FERNS.

Alsophila Australis.	Cyathea Dealbata.

FICUS ELASTICA.

A very ornamental plant, with very large, thick, glossy-green leaves; a fine decoration plant for the garden in summer, or the conservatory in winter. Price, $1.00 to $3.00 each.

FICUS REPENS.

A beautiful, fast-growing greenhouse plant, suitable for covering walls, pillars, etc., as it clings with the greatest tenacity to whatever it is planted against. Price, 25 cts. each; $2.50 per doz.

GARDENIA FLORIDA. — (Cape Jessamine.)

Plants with dark, glossy foliage; flowers pure white, blooming in spring, deliciously fragrant. Price, 50 cts. each.

GESNERA EXONIENSIS.

A valuable winter plant of great beauty. Flowers intense orange, in dense spikes; leaves a rich, velvety purple, studded with minute red hairs. Price, 50 cts. to $1.00 each.

GNAPHALIUMS.

Neat, white-leaved plants, suitable for baskets, vases, or narrow ribbon-lines.

G. Lanatum. Downy-white foliage, of a vigorous growth.
"   "   Variegatum. Same style of growth, with variegated leaves.
" Saundersonii. Very shrubby, growing about six inches high. Silvery white.
" Tomentosum. Narrow, lanceolate leaves, forming a bush ten to twelve inches high.

Price, 15 cts. each; $1.50 per doz.

GLOXINIAS.

Of these magnificent summer-blooming plants, we have a splendid collection, both of upright and pendant varieties. They are well worthy of a place in every collection. Price, 50 cts. each; $3.00 to $4.00 per doz.

GLADIOLUS.

No plant blooms so freely with little or no care as the Gladiolus. The varieties are now so numerous, and many of them so much resembling each other, that we do not give a descriptive list of varieties; no garden, however small, should be without them. We offer, this season, a choice collection of seedlings, surpassing many of the named varieties in beauty, and at one half the price. Price—Named varieties, 25 cts. to 50 cts. each; $3.00 per dozen; seedlings, 15 cts. each; $1.50 per dozen; $5.00 per hundred.

DOUBLE FLOWERING IVY-LEAVED GERANIUM. — (Kœnig Albert.)

A very free-growing variety, of trailing habit; flowers double, of a violet-rose color. A very useful decorative plant. Good for bouquets. Price, 10 cts. to 25 cts. each; $1.00 to $2.50 per dozen.

DOUBLE IVY-LEAVED GERANIUMS.

FINETTE. Blush-white, upper petals flushed with rose and feathered with dark crimson.

MDME. E. GALLE. Pure white, very double and large; the best.

M. DILBUS.

VISCOUNTESS CRANBROOK. White, shaded satin-rose. Very pretty; quite double.

Price, 25 cts. each.

VARIEGATED GERANIUM.

FREAK OF NATURE. An improvement on "Happy Thought," foliage much smaller; habit very dwarf and branching; centre of leaf pure white. Novel, distinct, and wonderfully attractive. Price, 50 cts. each.

GERANIUM. — (Happy Thought.)

This is a novel and interesting variety, differing from all others in the style of variegation, having a large yellow blotch in the centre of the leaf, with an outer band of green at the margin; flowers, rich magenta-rose; habit, dwarf. Price, 25 cts. each; $3.00 per dozen.

GERANIUMS. — (Zonale General Collection.)

ALICE SPENCER. White, pale-pink eye.

ATALANTA. Magenta, shaded purple.

AURORA. Rich salmon.

BRUTUS. Dark scarlet; very fine, single.

BURNS. Light scarlet, flowers large, and of perfect form.

CHIEFTAIN. Orange-scarlet.

CLEOPATRA. Flowers a brilliant shade of carmine-scarlet, trusses of fine form; very free-flowering.

COLONEL FIEVEL. Dark crimson-scarlet flower.

4

CORREGIO. Crimson, tinted purple; fine.

DANTE. Deep rose-pink, white centre.

DIANA. Orange-scarlet, large zone.

DUCHESS. Carmine-scarlet, large truss.

FANNY CATLIN. A beautiful variety. Color, soft rosy-salmon; large white eye; immense trusses; a strong grower.

GERTRUDE. Light-salmon.

GATHORN HARDY. Orange-scarlet, very fine.

GUINEA. Brilliant orange-yellow, very distinct.

GENERAL GRANT. Brilliant scarlet, fine bedding variety.

GLOW. Orange-scarlet, dwarf habit.

GLOBOSA MAJOR. A deep maroon-crimson flower, trusses of immense size and of perfectly globular form.

HEATHER BELL. Flowers a distinct blush-tinted or lilac-pink color, with a white blotch on upper petals, trusses large and globular.

HELEN DICK. Pale salmon-pink.

IRENE. Flowers a deep shade of magenta-purple, trusses large and produced in great abundance.

INCOMPARABLE. Orange-scarlet, white eye.

INTERNATIONAL. Rich crimson or crimson-scarlet, fine.

JEALOUSY. A very distinct color, orange-scarlet, of a yellowish hue.

JOYFUL. Light magenta, distinctly edged with orange-scarlet; flowers and truss large; fine habit.

LA CLAIN. A beautiful salmon.

LADY KENNEDY. Intense glowing scarlet, white eye, flowers single and perfect shape.

LAURA STRACHEN. Deep salmon, large trusses; free.

LOUIS VALLIOTT. Dark crimson-scarlet, dwarf habit.

MANFRED. Light orange-red, fine form.

MADONNA. Deep rosy-pink, white on upper petals; large.

MASTER CHRISTINE. Deep pink, very large trusses.

MADAME VOUCHER. Large, white, best.

MISS NELSON. Rosy-pink, fine bedding variety.

MRS. DURRELL. Beautiful carmine-pink, large trusses, and stand well above the foliage.

MRS. GLADSTONE. White, pale-salmon eye, large close trusses; flowers stand well above the foliage.

NEMESIS. A beautiful nosegay, color a very dark scarlet, trusses large.

NEW GUINEA. Orange-yellow, two shades lighter than Guinea; dwarf, compact habit; flowers well formed.

OLIVE CARR. Rosy-pink; flowers large, and good shape.

PLACCI. Scarlet, shaded magenta.

PHILAMONA. White, salmon eye.

PLACIDA. Deep pink, large white centre, single.

REV. A. NEWBY. White, pink eye.

RICEII. Bright scarlet, white eye.

SILVIO. Bright orange-scarlet, white eye.

SOPHIA BERKIN. Bright mottled salmon; trusses large and freely produced.

SUNBEAM. Brilliant scarlet, bright orange hue.

STRAIGHT AWAY. Maroon-crimson, upper petals scarlet. The petals overlap, and are of good substance; trusses large; dwarf habit.

SALMON RIENZI. A beautiful salmon self, of good size and substance.

TIPTOP. Crimson-suffused magenta; large white eye; flowers very large and finely formed; dwarf habit.

TITANIA (PEARSONS). Bright salmon, distinct, fine for winter blooming.

VESUVIUS. Scarlet dwarf.

WHEEL OF FORTUNE. Soft rosy scarlet; petals broad; finc-shaped flowers.

WM. PEARSONS. Fine dark scarlet.

Price, 20 cts. to 35 cts. each; $2.00 to $3.50 per doz., except where noted.

GERANIUMS.—(Double.)

This class forms an interesting and valuable addition, not only from their novelty, but from their great value as bedding plants. As the blossoms do not shed their petals, they remain a long time in flower, and are useful in various ways where single varieties are almost worthless.

ARETHUSA. Deep scarlet, fine.

BISHOP WOOD. Upper petals violet, lower orange-scarlet; abundant bloomer.

BATACLAN. Rich dark carmine, shaded purple.

C. H. WAGNER. Bright orange-scarlet, shaded purple.

CHAS. VOGHT. Deep pink shaded, semi-double.

DR. GENIGNEAU. Dark crimson, shaded magenta-rose, good truss.

DEPUTE BRICE. Large flowers, trusses large, petals the color of peach-blossoms, with tints of carmine; very novel.

DEPUTE BERLET. Beautiful shade of pink, deep tinge of violet.

DEPUTE VARROY. Enormous, rather flat trusses, color vivid lake, deep at the edges of petals, and shading to white at the centre; a magnificent shade.

DEPUTE D'ANCELON. A beautiful, well-shaped flower, color deep magenta-rose, large truss; a good grower.

DOLABLE. Brilliant scarlet, large truss.

EMILY LAXTON. Immense, vivid scarlet, velvety, very double, almost quilled.

HENRY BURHER. Fine orange-salmon.

HETERANTHE. Bright scarlet, semi-double, fine form.

HEROINE. The largest pip of any in the class; pure white; very fine.

J. C. RODBARD. Salmon-red.

LA BRONZE. Pale pink.

LOUIS POITIER. Deep salmon; guard petals blotch-white. Novel and beautiful.

M. PASTEUR. Deep crimson; flowers large and well formed.

MONSIEUR GELEIN LOUAGIE. Semi-double, very large; color, bright vermilion, shaded with salmon; dwarf habit.

MRS. MAMPION. Rose-pink, large flower.

MARIA BERTIER. White, shaded pale pink; large trusses.

MADAME THIBAULT. Very free-flowering, trusses large and fine, flowers very large and double, bright magenta-rose, upper petals marked with white.

MADAME GIBBARD. Brilliant pink.

SOUVENIR DE CARPEUX. Deep magenta; good habit.

SYLVIA. Rosy-pink.

SOUVENIR DE R. MASON. White.

THOS. GRIMES.

VESTA. Bright scarlet; small trusses; free flowering.

Price, 20 cts. to 35 cts. each; $2.00 to $3.50 per doz.

GERANIUMS. — (Gold, Silver, and Bronze Leaved.)

ALMA. Leaves deep green, silver margin; flowers bright scarlet.

BIJOU. Leaves round, silver margin; flowers bright scarlet.

CRYSTAL PALACE GEM. Leaves golden, green centre.

COUNTESS OF WARWICK. Sulphur-white, carmine zone.

COUNTESS OF CRAVEN.

EMPEROR OF BRAZIL. One of the best bronze varieties. 35 cts.

LADY CULLUM. Improvement on Mrs. Pollock. 50 cts.

MRS. POLLOCK. Leaves deep green, zone tinted with scarlet, yellowish-green margin.

MOUNTAIN OF SNOW. Deep green, white margin.

MARSHAL MCMAHON. Ground-color of the leaves golden-yellow, deep chocolate zone, extra fine variety. 35 cts.

MRS. J. CLUTTON. A fine silver tricolor.

MISS GORING. Golden-yellow leaves, margined with brilliant flame-colored tints; crimson-scarlet zone. 50 cts.

PRINCE SILVER WINGS. Green disk, bronze zone, tinted with carmine and white.

PRINCE OF TRICOLORS. Bright green disk, with crimson zone, interspersed with fiery scarlet.

SOCRATES.

WM. SANDY. Like Miss Goring in color, but of a stronger growth. 50 cts.

Price, 25 cts. each, except where noted; $2.25 per doz.

GERANIUMS. — (Hybrid Bedding and Scented.)

LADY PLYMOUTH. Variegated foliage, rose-scented.

MORGANII. Dark scarlet flowers.

PRINCE OF ORANGE. Delicious scent.

ROSE.

RADULA. Singular-cut foliage.

UNIQUE (Rollinson's.) Deep crimson-purple flowers,

WHITE UNIQUE. Light flowers, spotted.

Price, 20 cts. each; $2.25 per doz.

GERANIUMS. — (Ivy-Leaved.)

BRIDAL WREATH. Pure white.

LADY EDITH. Leaves green, flowers crimson, shaded purple.

L'ELEGANTE. Foliage bright green, with broad bands of pure white; flowers pure white; fine for baskets, etc.

WILLSII. A new hybrid of vigorous habit and compact growth, flowers violet-rose color.

WILLSII ROSEA. Like the last in habit, with delicate rosy-pink flowers.

Price, 20 cts. to 30 cts. each; $2.50 per doz.

GRAPTOPHYLLUM NORTONII.

A foliage plant of neat habit. The leaves are of ovate oblong form, about six inches long and two and a half broad; deep glossy green. The mid-ribs are at first light rose, and have on one or both sides a central blotch of light yellow. As the leaves approach maturity, the mid-ribs deepen to crimson, and the blotches becomes suffused with rose. Price, 50 cts. each.

HELIOTROPES.

A well-known and much-admired fragrant plant. Our collection contains the best of the light and dark varieties. Price, 10 cts. to 30 cts. each; $1.00 to $3.00 per doz.

HERBACEOUS PLANTS.

Herbaceous plants are the most valuable of all garden-flowers. They are not only hardy, but easy of culture, increasing rapidly, and give a succession of bloom from May to November.

We have added a large number of choice new and old varieties of this valuable class of plants to our stock this year, that are not enumerated in the following list:

ANEMONE FULGENS. Scarlet.

"   STELLATA FULGENS. Dazzling, fiery scarlet.

"   In variety.

AQUILEGIA (Columbine). Well-known flower, throwing up stems of flowers of various colors, two feet high.

BAPTISIA. Handsome spikes of blue, lupine-shaped flowers, in June and July.

DAISY (Bellis). See special description.

DIELYTRA (Bleeding Heart). A beautiful border-plant, with brilliant rosy, heart-shaped flowers.

DELPHINIUM. Four sorts.

CANTERBURY BELL. Large, showy, bell-shaped flowers of pure white, rose and purple; June.

FOX GLOVE (Digitalis). Showy, bell-shaped flowers on stems three or four feet high.

FRAXINELLA. Small spikes of white and reddish-purple flowers, in June.

FUNKIA JAPONICA ALBA. White.

"   "   Foliage variegated, light blue.

GLADIOLUS COLVILLI (The Bride). The finest white gladiolus among the early-flowering section; invaluable for market purposes, excellent for cut-flowers. It is quite hardy, and will grow in any light soil.

HELLEBORUS NIGER (the common Christmas rose). Greenish-white; fine for forcing.

HELLEBORUS NIGER MAXIMA (the giant-flowered). Flowers large, pure white, and freely produced.

HELLEBORUS ANTIQUORUM. Large bell-shaped flowers, sepals imbricated, pure white, free-flowering.

HELLEBORUS ABCHASICUS. Deep, rich purple.

"   ATRORUBENS. Light rose-purple; robust habit; free.

"   COLCHICUS. The young stems and foliage, rich purple; flowers, much imbricated, of a deep rich plum-color.

HELLEBORUS GUTTATUS. Pure white flowers, freely spotted with purple, one third up the petals, petals deep green; a very showy variety.

HELLEBORUS GUTTATUS LEICHTLINI. Similar to above, but the spots are larger and more distinct.

HELLEBORUS GUTTATUS SUB-PUNCTATUS. White.

"   LIVIDUS. Flowers a bright green; handsome foliage.

"   ORIENTALIS. Flowers large and globular, inside bright rose, outside deep purplish-red.

HELLEBORUS ORIENTALIS ELEGANS. Greenish-white, tinged with purple and rose.

HYACINTHS. In variety.

HOLLYHOCK. A fine collection of all colors.

IRIS. In variety.

IBERIS SEMPERVIRENS (Perennial Candytuft). Low-growing plants, with heads of white flowers.

LYCHNIS VISCARIA. Pinkish-red, double flowers.

LILY OF THE VALLEY. White.

MYOSOTIS (Forget-me-not). Blue.

NARCISSUS POETICUS ORNATUS. Flowers large, pure white, with a red cup.

NARCISSUS. In variety.

PANCRATUM. See special description.

PRIMULAS. In variety.

PHLOX SUBULATA (Moss Pink). Low-growing, covered in spring with pink flowers.

POLYANTHUS. In variety.

PHLOX. See special description.

SPIRÆA JAPONICA. Feathery white flowers, one foot high.

"   "   Fol. variegata.

"   PALMATA. The finest of the "meadow sweets"; large corymbs of crimson-purple flowers, three feet high.

SPIRÆA PLUMOSA.

"   In variety.

SEDUM. For rock-work. See special description.

SWEETWILLIAMS. In variety.

Vinca (Periwinkle). White and blue, for cemetery.
Wahlenbergia. Blue and white.
Yucca (Adam's Needles). Showy, cream-white flowers.

Price, 25 cts. to 50 cts. each; $2.50 to $5.00 per doz.

HERNIARIA GLABRA.

A neat, low-growing plant, with dark glossy-green leaves; a valuable plant for fancy bedding. Price, 15 cts. each; $1.50 per dozen.

HIBISCUS.

A showy and handsome plant, of rapid growth, forming large bushes. The foliage is a glossy green. The Hibiscus is valuable both for summer and winter blooming.

H. Cooperii Tricolor. Foliage beautifully mottled with pure white and pale rose-color; flowers bright crimson.
H. Double. Red.
"   "   Yellow.
"   "   Crimson-scarlet.

Price, 25 cts. to 50 cts. each; $2.50 to $4.00 per doz.

HYDRANGEA SPECIOSA.

A very conspicuous and showy variety, having very large leaves with a blotch of pure white through the centre of each. Price, 35 cts. each.

HYDRANGEA JAPONICA VARIEGATA.

Beautiful foliage of deep green, marked with white. Adapted to a rather shaded location in the garden, as the leaves burn when exposed to our hot mid-summer sun. Excellent for greenhouse decoration during the summer. Price, 25 cts. each; $2.50 per doz.

HYDRANGEA PANICULATA GRANDIFLORA.

A beautiful, hardy variety from Japan, with immense trusses of pure white flowers, changing to blush, and remaining in blossom from August until frost. Price, 50 cts. each.

HYDRANGEA OTAKSA.

A fine variety of this popular plant, producing immense panicles of rosy-carmine flowers when the plants are quite small. Price, 30 cts. each; $3.00 per doz.

HYDRANGEA.—(Thomas Hogg.)

One of the novelties, introduced from Japan. It belongs to the Hortense section of the family, but it is a far more free and abundant bloomer than any other kind. The flowers are of the purest white, and of very firm texture. The plants blossom when quite small and continue in bloom for a great length of time. A valuable plant for cemetery decoration. Having a large stock, we have put them at a very low price. Price, 25 cts. to $2.00 each.

IRIS.

The Iris are a very beautiful class of plants. They are of various shades of blue, purple, white, and yellow. Perfectly hardy, and require no attention after being planted.

JAPANESE MAPLES.

These Maples are among the most beautiful and interesting additions to our ornamental, deciduous trees or shrubs, that have been made within the past few years. They are low trees or shrubs, with erect stems, the branches more or less spreading, in the different varieties, and clothed with foliage that is developed in a greater variety of form and color than in any other species of deciduous trees known. Price, $1.00 each.

JASMINUM GRANDIFLORUM.

A beautiful winter-flowering plant. The flowers are pure white, very fragrant. Price, 30 cts. to 50 cts. each.

KLEINIA REPENS.

A very pretty succulent plant, with long, roundish, glaucous leaves. A very desirable basket-plant. Price, 25 cts. each.

LEUCOPHYTON BROWNII.

A very pretty variety of the white or silvery-leaved plants, for ribbon or carpet bedding. Price, 10 cts. each; $1.00 per doz.

LOBELIAS.

Plants of easy growth, well adapted for bedding, edging, rockeries, baskets, etc. Price, 10 cts. each; $1.00 per doz.

LOMARIA GIBBA.

A greenhouse Tree Fern, of the most elegant growth, adapted for every purpose, graceful in its feathery fronds, easy of culture in any shady wood earth; grows rapidly; adapted for baskets, vases, etc.; bears exposure to the sun like a Sago Palm, to which it has a strong resemblance. Price, 15 cts. to 50 cts. each; $1.50 to $4.00 per doz.

LYGODIUM SCANDENS. — (Japanese Climbing Fern.)

A climbing Fern of graceful, twining habit, often attaining a height of twenty feet, and is much used as a substitute for smilax in decorating. It is of easy culture, and is a handsome plant for vases or hanging baskets, as it can be used to climb or to droop as required. Price, 25 cts. each; $2.50 per doz.

MARANTA.

A beautiful ornamental-foliage plant, suitable for ferneries, baskets, vases, etc.

ARGYREA.	MICANS.	REGALIS.
BICOLOR.	MAKOYANA.	VAN DEN HECKEI.
LITZIANA.	PULCHELLA.	ZERINA.

Price, from 50 cts. to $2.50 each.

MAHERNIA ODORATA.

A very desirable greenhouse plant, blooming profusely during the spring months. The flowers are yellow, bell-shaped, and very fragrant. Price, 25 cts. each.

MYRSIPHYLLUM ASPARAGOIDES. — (Smilax.)

There is no plant in cultivation that surpasses this in the graceful beauty of its foliage, and its peculiar, wavy formation renders it one of the most valuable of all plants for vases and baskets, as it can be used either to climb or to droop, as required. The foliage is smooth and glossy; indespensable for bouquets, wreaths, etc. Price, 10 cts. each; $1.00 per doz.

MEYENIA ERECTA.

A beautiful greenhouse shrub, with light blue flowers; flowering profusely during the spring months. Price, 50 cts. each.

MAURANDIAS.

Beautiful climbing plants, of graceful, slender growth, producing an endless profusion of handsome white, pink, or purple gloxinia-shaped flowers throughout the season.

ALBA. White. BARCLAYANA. Purple. ROSEA. Rose.

Price, 10 cts. to 25 cts. each; $1.00 to $2.00 per doz.

MATICARIA INODORA. — (The Double May Weed.)

Is said to be of unusual attraction, and quite a new introduction; it is similar to a chrysanthemum; of easy cultivation, and produces its flowers in great abundance; valuable acquisition. Price, 15 cts. each; $1.50 per doz.

MACROLEPIA HERTA CRISTATA.

A beautiful fern, having a dwarf and spreading habit; fronds crested, the plunæ branched and subdivided near their extremities. The color is soft and

5

pleasing, and the much-divided pinnæ give the fronds a very elegant and attractive appearance. Price, 50 cts. each.

NERTERA DEPRESSA.

A neat little Alpine plant, with very small dark green leaves. It produces berries of brilliant orange-scarlet, forming a pleasant contrast with the foliage. A splendid plant for baskets, rock-work, etc. Price, 25 cts. each.

NIEREMBERGIA.

Very beautiful, neat-growing plants, of good habit and delicate foliage, with a profusion of pretty purple and white flowers from June to September. Well adapted for baskets, vases, etc. Price, 10 cts. to 25 cts. each; $1.00 to $2.25 per doz.

ORCHIDS.

As the prices of orchids vary considerably, according to the size of the plants, we think it best not to affix any price to this list; but we shall be pleased to give all information on application.

CALANTHE VEITCHII.
"   VESTATA.
CŒLOGNE CRISTATA.
CYPREPEDIUMS IN VAR.
DENDROBIUM BIGIBBUM.
"   BENSONII.
"   CRASSINODE.
"   CRYSTALLINUM.
"   DENSIFLORUM.
"   DEVONIANUM.
"   FORMOSUM.
"   FORMOSUM GIGANTEUM.
"   NOBILIS.
"   PIERARDI.
"   THYRSIFLORUM.
"   WARDIANUM.
EPIDENDRON VITELLINUM.
LÆLIA ALBIDA.

LÆLIA ANCEPS.
"   AUTUMNALIS.
"   PURPURATA.
ODONTOGLOSSUM ALEXANDRA.
"   CIRRHOSUM.
"   GRANDE.
"   PULCHELLUM.
"   VEXILARIUM.
ONCIDIUM.
"   FORBESII.
"   INCURVUM.
"   TIGRINUM.
"   VARICOSUM.
PERESTERIA ELATA.
PHAJUS GRANDIFLORA.
PILUMNIA FRAGRANS.
ZYGOPETALUM MAKAYII.

PANSIES.

We have a very choice collection, raised from the seed of the best English and American growers. Price, 10 cts. to 15 cts. each; $1.00 to $1.50 per doz.

PÆONIES.

One of the showiest and most desirable of hardy, ornamental plants, with beautiful, sweet-scented flowers. Their easy culture recommends them to every one who has a garden. Price, 25 cts. to 50 cts. each; $2.50 to $4.00 per dozen.

PALMS.

As a decorative plant they stand unrivalled. They are easily cultivated, and require but little attention; of graceful and stately form, they impart a rich, sub-tropical appearance wherever used.

BRAHEA FILAMENTOSA. 50 cts. each.

CARYOTA URENS. Wine palm, small plants; 75 cts. each.

CYCAS REVOLUTA. The sage palm, very choice; $1.00 to $5.00 each.

DION EDULE.

LATANIA BORBONICA. Beautiful, with immense fan-shaped leaves; $1.00 to $10.00 each.

PHŒNIX DACTYLIFERA. Date palm.

PTYCHOSPERMA. A fine palm.

PASSIFLORA. — (Passion Vine.)

A beautiful free-growing climber, with handsome dark-green leaves and showy flowers.

PASSIFLORA CÆRULEA. Blue.

DECASIANA. Pale blue-red and purple.

" VARIEGATA. Same color, with variegated leaves.

PEPEROMIA.

Beautiful low-growing plants, with variegated foliage. A choice plant for ferneries, etc.

MACULOSA; 25 cts. each.   VERSCHAFFELTII; 25 cts. each.

SHOW AND FANCY PELARGONIUMS.

These Pelargoniums are remarkable for their large, showy forms, easily grown, flowering profusely, and present in their varied tints and colors what few other plants possess.

ADANSON. Deep purple crimson, violet centre.

ALMA.

ALEXINA.

BRIDESMAID. Pale lavender, edged with white.

BERTHA.

BLANCHE FLEUR. Nearly white.

BEACON.

BALTIC. Flame color, maroon spot.

CHARLES KEEN. White centre, rich lower petals, maroon top, shaded with carmine.

CORSAIR. Light purple, with pure white centre, petals blotched with maroon crimson.

CHIEFTAIN. Pink, blotched and spotted with dark purple.

CORONET. Crimson-rose, dark crimson spots.

CRIMSON KING. Crimson, blotched maroon, very showy.

DR. ANDRY.

DR. MASTERS. Rosy crimson.

DESDEMONA. Pure white, deep maroon blotch.

DUCHESS OF BEDFORD. Pure white, slightly spotted on upper petals.

EVENING STAR. Purple, profuse bloomer.

FRIEND EDE. Scarlet, maroon spots.

GLORY OF AMERICA. Blush-pink, dark spots.

GRANDES. Carmine, light centre; maroon spot on every petal; good habit, and free-flowering.

HER MAJESTY.

HARLEQUIN. Carmine, dark spot, white throat.

IMPERATRICE EUGENIE. French-white, rich, dark spots.

JULIET. Light ground, upper petals bright purple, lower blotched with crimson.

LE NEGRO.

MARY HOYLE. Orange-rose, white centre.

MRS. L. FLOYD. Bright scarlet, profuse bloomer.

MERMERIS. Dark rich scarlet, bright centre; maroon spot; very free.

MARIA.

PERSEVERANCE.

PRINCE OF PELARGONIUMS. Vermilion-scarlet, blush-white centre, veined violet, upper petals flushed with crimson.

QUEEN VICTORIA. Rich vermilion, broadly margined with white-maroon spot.

REINE HORTENSE.

RESPLENDENT.

ROSA MUNDA.

ROSE CELESTIAL. Delicate pink, deep rose blotch.

SIR J. PAXTON. Very dark purple, free bloomer.

SCARLET GEM. Rosy-scarlet, light centre; fringed; very beautiful.

TRIOMPHE DE ST. MAND. Rich deep crimson, large trusses; dwarf; strong habit; free.

TORCH. Dark crimson, margined with rose, crimson spots.

VESTAL. Pure white, dark spots on upper petals.

WM. BULL. Crimson-scarlet, dark-spotted, free bloomer, very fine.

WHITSTONE HERN.

Price, 25 cts. to 50 cts. each; $3.00 to $5.00 per doz.

FLORIST'S OR PAISLEY PINKS.

In variety. Price, 25 cts. each; $2.50 to $4.50 per doz.

PILEA MUSCOSA AND SERPÆFOLIA.

Very pretty, neat-growing plants, with slender, frond-like leaves; desirable for baskets, vases, etc. Price, 25 cts. each; $2.25 per doz.

PETUNIAS.—(Single, Striped, and Blotched.)

Our stock of Petunias has been much improved by saving only from the best flowers, so that we now have a superior strain, which for beautiful markings and solidity of flowers can scarcely be equalled. Price 15 cts. each; $1.00 per doz. ; $8.00 per hundred.

PHLOX.

A fine collection of this most interesting of our hardy, herbaceous perennial plants. They are of the easiest culture, and embrace every color, from purest white to deepest crimson; well adapted for city gardens or shrubberies, as they will grow either in shade or sunshine.

ADVANCER. Purplish-crimson, dark eye.
AUGUSTA S. RIVIERE. Bright vermilion, crimson eye.
AUGUSTA RIVIERE. Pure white, purple-crimson eye.
ARGOSY. Crimson, centre bright, dark eye.
BRILLIANT. Dark rosy pink, large crimson eye.
CALYPSE. Violet-crimson, large.
CHAS. WAGNER. Deep bright crimson, maroon eye.
COCCINEA. Crimson-scarlet, fine.
COUNTESS OF BREADALBANE. Dark purple-crimson.
CYPRUS. Salmon-pink, dark eye.
DEFIANCE. Violet-crimson, bright.
DR. BOURDEVALL. Beautiful carmine-rose, large.
DUKE OF LANCASTER. Deep rose, dark centre.
ERNEST BENARY. Pale rose, deep eye.
EDITH. Pure white, violet-purple eye.
GLOIRE DE NEUILLY. Deep vermilion, shaded crimson, dark eye.
J. K. LORD. Vermilion, shaded crimson, bright eye.
JOHN FORBES. Clear pale rose, crimson eye, very large and fine.
JEANNE D'ARC. Snow-white, fine.
LOUIS VAN HOUTTE. Vermilion, maroon eye.
LARINA. Pure white.
LOTHAIR. Deep crimson, shaded scarlet, carmine centre.
MDME. FORBOIS. Blush, large violet, crimson eye.
MDME. BRANDTT. White, bright eye.
MARIA LAISON. Magenta, shaded violet, dark eye.
MR. WARE. Purplish-crimson.
M. DUBUE. Salmon-pink, crimson eye.
POMME DE SIDON. Dark violet-crimson, fine.
RESPLENDENT. Magenta, shaded scarlet, fine.
RUBY. Reddish-purple, fine.
S. S. WARE. Dark-lilac, light centre.
SPLENDENS. Magenta-rose.
SOUVENIR. Bright vermilion, dark eye.
VIRGO MARIA. Fine white.
YORK AND LANCASTER. Light rose, purple-crimson eye.

Price, 25 cts. each ; $2.50 per doz.

PANICUM VARIEGATUM.

A very pretty, neat-growing, variegated grass, of drooping habit. The color of the leaves is dark green, rose, and white; desirable for baskets, vases, etc. Price, 15 cts. each; $1.50 per doz.

PANDANUS GRAMINIFOLIUS.

A fine, dwarf-growing variety, of drooping habit; with deep green, narrow leaves, serrated on the edges. Price, 50 cts. each.

PANDANUS UTILIS.

A beautiful plant, with long, recurved, glossy green leaves; well adapted for centres of vases, baskets, etc. Price, 50 cts. to $1.00 each.

PANDANUS VEITCHII.

A most beautiful variety, leaves light green in color, with stripes and bands of pure white; most graceful habit. Price, $1.00 to $5.00 each.

PHORMIUM TENAX VARIEGATUM.

A variegated form of the well-known New Zealand Flax. The leaves, which are often six feet long, are of a dark green color, distinctly marked with broad stripes of yellow, more than half the leaf being frequently of the latter color. It is very ornamental for conservatory and summer flower-garden decoration. Price, $3.00.

PTERIS SCABERULA.

A beautiful fern, with feathery fronds, of rapid-spreading growth. Choice for ferneries, baskets, and valuable to the florist for its elegant foliage. Price, 25 cts. each; $2.50 per doz.

PERISTROPHE ANGUSTIFOLIA AUREA.

A beautiful plant, with brilliant golden foliage, striped with green; excellent for baskets and vases. The flowers are a delicate violet color; free bloomer. Price, 20 cts. each; $2.50 per doz.

POINSETTIA PULCHERRIMA.

A magnificent plant, producing, about Christmas, the most gorgeous bracts of rich vermilion, which often measure one foot in diameter, and remain in perfection several weeks. Price, 50 cts. to $1.00 each.

REINECKIA CARNEA VARIEGATA.

A beautiful variegated grass-like plant of dwarf habit; an excellent plant for aquariums, ferneries, etc. Price, 25 cts. each.

RHYNCHOSPERMUM JASMINOIDES.

A beautiful greenhouse climber; flowers pure white and very fragrant; a valuable winter-blooming plant. Price, 25 cts. each.

RHODODENDRONS.

We would call the attention of our patrons to our superior stock of hardy Rhododendrons. These beautiful plants are yet but little cultivated, undoubtedly in a great degree from the general impression that they are difficult to manage, requiring special care and what is called peat soil, and it was at one time believed that they would not thrive in any other. Experience, however, proves the contrary, and it is now found that Rhododendrons thrive in almost any soil that does not contain lime. Strong plants, well set with buds, from two to three feet high, $2.00 to $5.00 each.

BENNETT'S HYBRID TEA ROSES.

BEAUTY OF STAPLEFORD. Pale pinkish rose, shaded darker in centre; good form.

DUCHESS OF CONNAUGHT. Delicate silvery rose, with a bright salmon centre, sweetly scented.

DUCHESS OF WESTMINSTER. Bright cerise, very large.

HON. GEO. BANCROFT. Bright rosy crimson, edged with bright red, large and of fine form.

JEAN SISLEY. Rosy lilac, turning to bright pink in centre, large and full.

MICHAEL SAUNDERS. Bronzey-pink, sweet-scented, very large, and of good form.

NANCY LEE. Bright satin rose, of good form.

PEARL. Flesh color, not large, but well formed.

VISCOUNTESS FALMOUTH. Very delicate pinkish rose, tinged with darker pink, large and of good form.

Price, 25 to 50 cts. each; $2.50 to $4.50 per dozen.

CHOICE TEA ROSES.

ALINE SISLEY. Deep purplish rose, shaded violet-red, medium size, fine form, delicious fragrance.

CATHERINE MERRMET. Delicate flesh-color, large, fine show-flower.

COMTESSE RIZA DU PARC. Beautiful metallic rose, changing to pink, large, full, and good form.

CORNELIA COOK. Creamy white, very large.

DUCHESS OF EDINBURGH. Crimson, most desirable color, fine form.

MDME. BRAVY. Cream, centre blush, beautiful shape.

MDME. CAMILLE. Delicate rose, violet shade, veined flowers, very large and full; a fine rose.

MDME. SURTOT. White, large, full, and fairly formed.

MDME. WILLEMOZ. Creamy white, centre tinted with fawn, petals very thick and fairly formed.

PEARL DES JARDINS. Straw-color, very fine, one of the best.

PEARL DE LYONS. Deep form and apricot, a beautiful large rose.

Price 25 to 50 cts. each; $2.50 to $4.50 per dozen.

BOURBON ROSES.—(Pillar Roses.)

BARRONNE DE NOIRMONT. Fresh rosy pink, petals of good substance, flowers large, beautifully formed, delicious violet fragrance.

MICHAEL BONNET. Fresh rose, flowers full and well formed, a fine rose.

SIR JOSEPH PAXTON. Bright rosy crimson, a free-blooming, handsome kind for pillars.

Price, 25 to 50 cts. each.

NOISETTE ROSES.—(Climbing Roses.)

BOUQUET D'OR. Deep yellow, centre copper-color, large and full.

MDME. CAROLINE KUSTER. Centre canary yellow, outer petals pale lemon, flowers large, globular, a fine rose.

MARECHAL NIEL. A rich brilliant yellow, too well known to need any description.

Price, 25 to 50 cts. each.

MONTHLY ROSES.

AGRIPPINA. Brilliant crimson.
BON SILENE. Rich, deep pink.
CANARY. Yellow.
GLOIRE DE DIJON. Fawn and rose.
HERMOSA. Delicate rose.
LAMARQUE. Pure white, in clusters.
LOUIS PHILIPPE. Light crimson.
MALMAISON. Blush, fine flower.

MARECHAL NIEL. Beautiful, deep yellow, large, full, and very sweet.
MADAME BOSANQUET. Pale, flesh color.
NIPHETOS. Pure white, extra.
PAULINE LABONTE. Light blush.
SAFRANO. Orange-yellow, splendid in bud, free bloomer.
WHITE TEA. White.
YELLOW TEA.

Price, 15 cts. to 50 cts. each; $1.50 to $5.00 per dozen.

HYBRID PERPETUAL ROSES.

Most of our roses are grown in pots, thereby avoiding all risk of transplanting, and ensuring to the purchaser plants that are alive and the certainty of their growth. This class of roses are entirely hardy, and bloom in spring and fall; they are greatly prized for their large, full-shaped flowers, which they produce in abundance.

Purchasers leaving the selection to us, will get a fine assortment of varieties and first-class plants. Price 50 cts. to $1.00 each; $5.00 to $10.00 per dozen.

PRAIRIE ROSES.

BALTIMORE BELLE.   ELEGANS.   QUEEN OF THE PRAIRIES.

Price, 50 cts. each.

MOSS ROSES.

Price, 50 cts. to $1.00 each; $5.00 per doz.

SALVIAS.

BETHELI. A splendid variety, excellent for autumn and winter blooming; dwarf habit, with beautiful pea-green foliage, and produces large spikes of the brightest rose.

PITCHERII. Another of the hidden genus; color most intense blue, flowers small in comparison with other varieties, more like a lobelia, but blooms very persistently, and set very close; for cut flowers all winter, it is most valuable.

SPLENDANS BRUANTI. This is much the best of all the brilliant scarlets for autumn and winter blooming; habit dwarf; foliage, pea-green; very free-flowering; spikes large and of a most intense color.

Price, 50 cts. each.

SONERILA HENDERSONII.

An exceedingly beautiful stove plant, of dwarf habit. The leaves are an olive green, densely marked with silvery white pearl-like spots. Price, 25 cts. each.

SONERILA HENDERSONII ARGENTEA.

Another handsome variety, with silvery leaves; in fact, the entire plant seems to be made of frosted silver. Price, 25 cts. each.

SUMMER AND OTHER CLIMBERS.

	Each.
AMPELOPSIS VEITCHII. Small variety of the Virginia Creeper	$0 25
BIGNONIA RADICANS.	50
COBÆA SCANDENS	25
" " VARIEGATA.	50
CLEMATIS. In variety	50
CHINESE WISTARIA.	50
HONEYSUCKLE. Monthly, blooms all summer, very fragrant	25
" Scarlet trumpet, monthly, blooms all summer, very showy	25
IVY. English	$0 25 to 1 00
" Variegated	35 to 1 00
LONICERA AUREA RETICULATA. A beautiful hardy climber, leaves being bright green, netted all over with yellow veins	25
MAURANDIA. Blue and white.	15
MADEIRA VINE.	15
PASSION FLOWER. In varieties	25
TROPÆOLUM. In variety.	10 to 15
VINCA MINOR.	15 to 50
" MAJOR. Variegated	15 to 50
VIRGINIA CREEPER.	25 to 50

6

SUMMER FLOWERING BULBS.

Each.

LILIUM AURATUM, or Golden-Banded Lily, universally acknowledged
    to be the finest of all lilies $0 50
    " LANCIFOLIUM ALBUM. Pure white 50
    "   "   RUBRUM. White, spotted with crimson . . 50
    "   "   ROSEUM. White, spotted with rose . . . 50
    " LONGIFLORUM 25
    " CANDIDUM. Is the well-known white garden lily, fragrant . 25
TRITOMAS 25
TUBEROSE. One of the choicest summer-flowering bulbs; the flowers
    are white and very fragrant; indispensable for making bouquets:
Started in pots 3 00
Dry Roots 1 25

SANTOLINA CHAMÆCYPARISSUS.

One of the best plants for edgings or ribbon lines, growing about a foot
high. Price, 15 cts. each; $1.50 per dozen.

SANCHEZIA NOBILIS VARIEGATA.

A very handsome hot-house plant, with leaves from twelve to fifteen inches
long, beautifully veined, and marbled with deep, golden yellow. Price, 30
cts. to 50 cts. each.

SEDUM.—(Stone Crop.)

The Sedums belong to the same family as the Echeverias and Sempervivums.
Being succulent plants, they are among the most valuable plants we have for
hanging-baskets, vases, and rock-work. They are also extensively used for
the edging of beds, or forming outer marginal lines, being dwarf and compact.
The flowers of some of the varieties are very beautiful. They are of the
easiest culture, either for garden or parlor decoration.

SEMPERVIVUMS.—(House Leek.)

A succulent genus of plants allied to the Sedums. They are unsurpassed
for rock ornamentation, many of them being hardy.

SIBTHORPIA EUROPEA.

A neat little plant of creeping habit, very effective when grown in baskets,
giving it a graceful appearance; an excellent plant for rock-work, moss-
baskets, etc. Price, 15 cts. each; $1.50 per doz.

STATICE HALFORDI.

One of the most beautiful greenhouse plants, with large, broad, light
green foliage; flowers bright blue and white, very showy, and remain in
perfection two or three months if kept in a cool house. Price, $1.00 to
$5.00 each.

TORENIA ASIATICA.

A very pretty summer plant, with small blue flowers, gloxinia-like shaped; well adapted for summer baskets, vases, etc. Price, 25 cts. each; $2.50 per doz.

TRITOMAS.

Splendid, half-hardy border-plants, flowering from July to October, and produce long spikes of orange-red and scarlet tubular flowers, each raceme from one to two feet in length. They are well adapted for forming large, effective groups and beds, in which the numerous terminal flame-colored blossoms have a fine effect. Price, 30 cts. each; $3.00 per doz.

TUBEROSE.

One of the choicest summer-flowering bulbs; the flowers are white and very fragrant; indispensable for making bouquets. Price, started in pots, $3.00 per doz.; dry roots, $1.25 per doz.

TUBEROSE. — (Pearl.)

This variety is generally conceded superior to the old variety; the flowers are of double the size, and imbricated like a rose, of dwarf habit, growing only two feet in height. Price, dry roots, $1.25 per doz.; started in pots, $3.00 per doz.

VINCA VARIEGATA.

A beautiful fast-growing plant, with bright green leaves, edged with yellow, and flowers of deep blue. Well adapted for baskets, rock-work, or vases. Price, 15 cts. to 50 cts. each; $1.00 to $4.50 per doz.

SWEET-SCENTED VIOLET. — (Marie Louise.)

This is undoubtedly the best violet offered for years. It surpasses all other varieties in the profusion of its flowers. In color it is darker than the Neapolitan Violet, double its size, and quite as fragrant. Price, 25 cts. each.

VERBENAS. — (General Collection.)

The following varieties of this popular plant we have selected from our large collection, the selections being made with great care, with a view to having the greatest variety of color combined with the best bedding qualities.

ALBA PERFECTA. White, very fragrant.
ARAN. Large, purple-magenta.
AIMEE. Pale mauve, yellow eye.
ANTLER. Crimson, large white eye.
BEAUTY OF OXFORD. Large, pink, fine.
BEAUTY OF SHERWOOD.
BRILLIANT DE VAISE. Crimson-scarlet.
BLUE CHAMPION. Indigo-blue.

COLOSSUS. Crimson, violet eye.
COMET. Crimson-scarlet, white eye, fine.
CELESTIAL BLUE. Extra fine, blue.
CANOBIE. Carmine, shaded violet, white eye.
CUBA. White, rose-pink stripe, distinct.
DAYBREAK. Blush-white.
DAZZLE. Blood-red, black markings.
FAUST. Rosy-salmon, shading to pink.
GOVERNOR TILDEN.
GENERAL CUSTER. Rich scarlet, dark eye.
GIANT. Rich scarlet, yellow eye.
HONOR. Blush, dark centre.
JEWEL.
JUBILEE.
LONDON PRIDE. Large, claret-color.
MARIANA. Rosy-carmine, yellow eye.
MAROON. Dark maroon.
MATTIE. Pale rose, white eye.
METEOR. Light red, fine.
MIKADO. Deep vermillion, shaded violet.
MISS ARTHUR. Dazzling scarlet.
MONTANA. White, striped carmine.
MOZART. Splashed scarlet and white.
MRS. KELLEY. Magenta, white eye.
MRS. WOODRUFF. Bright scarlet — the best scarlet.
NEGRO. Black, extra fine.
NEMESIS. Brilliant scarlet, white eye.
NIOBE. Large, pure white.
PROFUSION. Large blush, extra.
PASHA. Maroon.
RACCOON. Violet-purple.
ROSEA ALBA. Light pink, white eye.
ROSE QUEEN.
ROVER. Rich, dark maroon.
SCARLET CIRCLE. Dazzling scarlet, large white eye.
SYLPH. Pure white.
WATERLOO. Crimson, maroon centre.
WILLIE. Violet-crimson, yellow centre.
WHITE BEAUTY. White, large and fine.
WHITE BEDDER. White, large and full.
ZARA. White, splashed-purple.
ZENOBIA. Purple, large white eye.

Price, 75 cts. per doz.; $5.00 per hundred.

BASKET PLANTS.

ACHYRANTHUS LINDENII.
ACORUS GRAMINEUS. Variegated.
ALTERNANTHERAS. Four varieties.

ALYSSUM. Variegated.
BALM. Silver.
CALADIUMS. Varieties.

CENTAUREA GYMNOCARPA. Silver-gray foliage.

CINERARIA MARITIMA.

CISSUS DISCOLOR.

COLEUS. Varieties.

GNAPHALIUM LANATUM. Downy-white foliage.

IVIES. Of sorts.

LINERIA.

LOBELIAS. In variety.

LONICERA. Variegated Chinese Honeysuckle.

LYCOPODIUM. Of sorts.

MAURANDIAS. Two varieties.

PANICUM VARIEGATUM. A beautiful variegated grass.

CORONILLA GLAUCA. Variegated.

CUPHEA. Cigar-plant.

DRACÆNAS. Red and green.

EUONYMUS. Variegated.

FERNS. In variety.

FICUS REPENS.

GERMAN IVY.

PELARGONIUMS. Ivy-leaved.

PEPEROMIA MACULOSA.

SAXIFRAGA SARMANTOSA.

SEDUM CARNEUM VARIEGATUM.

TORENIA ASIATICA.

TRADESCANTIA. Two varieties.

TROPÆOLUM. Of sorts.

VINCA MINOR.

" MAJOR VARIEGATA.

Price, 10 cts. to 25 cts. each; $1.00 to $2.50 per doz.

MISCELLANEOUS BEDDING PLANTS.

	Per Doz.
ACHYRANTHUS. In variety	. $1 00
AGERATUM MEXICANUM. Best variety	. 1 00
ALOYSIA. Lemon Verbena	. 1 50
ALTERNANTHERA. Four sorts	. 1 00
ALYSSUM. Variegated-leaved	. 1 00
ANTIRRHINUMS. In variety	. 1 00
ASTERS. A fine assortment	. 50
BALSAMS. Camelia-flowered	. 50
BASKET PLANTS. In variety	. 1 50
BOUVARDIAS. In variety	. 2 50
CALCEOLARIAS. In variety	. 1 50
CARNATIONS. Monthly	. 1 50
CENTAUREAS. Silver foliage	. 1 50
CHRYSANTHEMUMS. Fifty sorts	. 1 50
CINERARIA MARITIMA. Fine silver-leaved plant	. 1 00
COLEUS. In variety	. 1 00
DAHLIAS. Fifty sorts	. 2 00
DAISY. Spotted leaves	. 1 00
DAISIES. Six sorts	. 1 00
FEVERFEW	. 75
" Golden Feather	. 65
FUCHSIAS. Best bedding varieties	. 2 50
GERANIUMS. Double	. 2 00
" Ivy-leaved	. 2 00
" Scented	. 2 00
" Variegated	. 2 00
" Zonale varieties	. 2 00
GNAPHALIUM LANATUM. Downy-white foliage	. 1 25

HELIOTROPE. Best bedding varieties   . $1 00
LANTANAS. In variety 1 50
LOBELIA. In variety   75
MADEIRA VINE   50
MAURANDIAS. Two varieties .   .   . 1 25
NIEREMBERGIA. In variety   .   .   . 1 25
PANSIES 1 00
    " Select seedings 1 50
PETUNIAS. Double 2 00
    "   Fine, single, blotched .   . 1 00
PHLOX. Fine variety 2 50
SALVIA SPLENDENS 1 00
STOCKS. Bedding 1 00
TROPÆOLUMS. Varieties   .   .   .   50
VERBENAS. With names   .   .   .   75
ZINNIA. Extra, double .   .   .   ●   50

SHRUBS.

ALMOND.
ALTHÆAS. In variety.
AZALEAS. In variety.
CALYCANTHUS.
CORCHROUS.
    "   Silver-variegated leaved.
DEUTZIA CRENATA FLORA PLENA. Flowers double, white, tinged with rose.
    "   GRACILIS. Flowers pure white.
    "   SCABRA. Profuse white flowers.
FORSYTHIA VIRIDISSIMA. Flowers bright yellow, very early in spring.
FRINGE TREE.
HAWTHORN. Double, pink and white.
HONEYSUCKLE, TARTARIAN. Red and white.
    "   SCARLET TRUMPET. Monthly, blooms all summer, very
    showy.
HONEYSUCKLE, YELLOW TRUMPET. Very fragrant.
LILAC. Common purple.
    "   Common white.
    "   Persian.
MAGNOLIA TRIPETALA. Fine, large leaves, white flowers.
MAHONIA AQUIFOLIA. Flowers yellow, early spring.
PURPLE FRINGE, or Smoke Tree.
PYRUS JAPONICA. Flowers bright scarlet, early spring.
RHODODENDRONS. In variety; $1.50 to $3.00 each.
SPIRÆA PRUNIFOLIA. Flowers white, blossoms in May.
    "   REEVESII. Round clusters of white flowers bloom in May.
SYRINGA, or Mock Orange. Three varieties.
TREE BOX.
VIBURNUM.

WEIGELIA ALBA. Flowers white, changing to a light, delicate blush.
    "   ROSEA. Rose-colored flowers, blooms in May.
    "   VARIEGATED-LEAVED. Leaves bordered with yellowish-white, flowers pink.
WISTARIA SINENSIS. (Chinese.)

Price, 50 cts. to $1.00 each.

WEEPING OR DROOPING DECIDUOUS TREES.

ASH. European weeping; the common well-known sort; one of the finest lawn and arbor trees.
    " Mountain weeping; a beautiful French variety, of rapid growth and decidedly pendulous.
BEECH. Weeping; a very graceful tree.
BIRCH. Cut-leaved, weeping.
    "   European weeping.
CHERRY. Weeping.
ELM. Weeping; an English variety, with smooth, glossy leaves.
POPLAR. Weeping.
WILLOW. American weeping.
    "   Kilmarnock weeping; one of the finest of this class of trees.
    "   Rosemary-leaved; branches feathery, with silver foliage.

Price, $1.00 to $5.00 each.

SHADE TREES.

A superior stock of Shade Trees of the kinds herein mentioned: Rock Maple, Horse Chestnut, Elm, Linden, Silver Maple, English Elm, Dwarf Horse Chestnut, and Magnolias. The Maples are unusually fine. Price, 50 cts. to $3.00 each.

FRUIT TREES. — (All the leading kinds.)

PEARS. Standard, 75 cts. to $1.00 each.
DWARF PEARS. 50 cts. to 75 cts. each.
PEACHES. 25 cts. to 50 cts. each.

EXOTIC GRAPES FOR VINERIES.

All the leading kinds. Price, one-year-old plants, 50 cts.; two-year-old plants, $1.50 each.

SMALL FRUITS.

CURRANTS. Best varieties, 15 cts. each; $1.50 per doz.
RASPBERRIES. 15 cts. each; $1.50 per doz.
STRAWBERRIES.
HARDY GRAPES. All the leading kinds. Strong plants, one-year-old, 25 cts. to $1.00 each; two-year-old, and extra vines, 50 cts. to $2.00 each.

HEDGE PLANTS.—(Evergreen. Stone.)

The idea of planting hedges for use and ornament, and screens for the protection of orchards, farms, and gardens, is a practical one, and rapidly becoming appreciated. Among the trees adapted to ornamental hedges the American Arbor Vitæ and the Norway Spruce take the first place. By using medium-sized plants a hedge can be made as cheaply as a good board fence can be built, and then, with very little care, it is becoming every year more and more "a thing of beauty." We all know that such hedges constitute the principal attraction in our best-kept places.

AMERICAN ARBOR VITÆ. $8.00 to $25.00 per hundred.
NORWAY SPRUCE. $25.00 per hundred.

DECIDUOUS.

PRIVET. One year, $5.00 per hundred.
JAPAN QUINCE. $15.00 per hundred.
DWARF BOX. For edging.

CHOICE CONIFERS.

ARBOR VITÆ. Siberian, the best of all the genus for this country, exceedingly hardy, keeping color well in winter, growth compact and pyramidal, makes an elegant tree; 50 cts. to $1.00 each.

ARBOR VITÆ. American Golden, beautifully marked with golden-yellow, very hardy; $1.00 each.

ARBOR VITÆ. Reidii, a dwarf bush of a beautiful shade of green, very dense and perfect form; 50 cts. to $2.00 each.

BIOTA ELEGANTISSIMA. End of young branches tipped with golden-yellow; $1.50 each.

JUNIPER, IRISH. A tapering, pretty little tree, $1.00.
    "   SWEDISH. A small-sized, handsome, pyramidal tree, with bluish-green foliage; $1.00.

JUNIPER, CHINENSIS. A small tree or shrub, with spreading branches; $1.00.
    "   COMMUNIS. A handsome, compact, small tree.

PINES. Austrian; a remarkably robust, hardy tree, growth rapid.
    "   Nordmanniana; this is a symmetrical and imposing tree; $2.00.
    "   Pichta; a medium-sized tree, quite compact and conical, and having very rich, dark foliage; $2.00.
    "   Cembra; a handsome and distinct European species, of a compact, conical form, foliage short and silvery; $1.00.

RETINOSPORA OBTUSA. A new evergreen tree from Japan, with beautiful light green foliage; $1.00.

RETINOSPORA PISIFERA. A small tree, with numerous delicate branches and feathery foliage; $1.00.

RETINOSPORA PLUMOSA. A variety with fine, short branches and small leaves; $1.00.

RETINOSPORA PLUMOSA AUREA VARIEGATA A beautiful variety, similar in habit to Plumosa, with foliage of a rich golden-yellow; small plants, 50 cts. each; large plants, $2.00.

MAHONIA AQUIFOLIA.

A beautiful evergreen shrub, yellow flowers, early spring, very hardy. Price, 50 cts. to $1.00 each.

KITCHEN GARDEN ROOTS, PLANTS, ETC.

ASPARAGUS. $1.00 per hundred. RHUBARB. 25 cts. each; $2.50 per doz.

PLANTS

Of Cabbage, Celery, Cauliflower, Egg Plant, Peppers, and Tomatoes, can be had in May or June.

NEW SINGLE DAHLIA.— (*See p. 4.*)

INDEX.